3 BcB8855

DFG

Analyses of Hazardous
Substances in Air

 WILEY-VCH

DFG Deutsche Forschungsgemeinschaft

Analyses of Hazardous Substances in Air

Volume 3

edited by Antonius Kettrup
Working Group Analytical Chemistry

Commission for the Investigation of Health
Hazards of Chemical Compounds in the
Work Area
(Chairman: Helmut Greim)

WILEY-VCH

Weinheim · New York · Chichester · Brisbane · Singapore · Toronto

Prof. Dr. Helmut Greim
Senatskommission
zur Prüfung gesundheitsschädlicher Arbeitsstoffe
der Deutschen Forschungsgemeinschaft
GSF Forschungszentrum für Umwelt und Gesundheit
Institut für Toxikologie
Ingolstädter Landstraße 1
D-85764 Oberschleißheim

Prof. Dr. Antonius Kettrup
GSF Forschungszentrum für Umwelt und Gesundheit
Institut für Ökologische Chemie
Ingolstädter Landstraße 1
D-85764 Oberschleißheim

Translators: Dr. Richard H. Brown, Mrs. Julia Handwerker-Sharman,
Dr. Karl-Heinz Ohrbach, Dr. Ann E. Wild

Deutsche Bibliothek Cataloguing-in-Publication Data:

Analyses of hazardous substances in air / DFG, Dt. Forschungsgemeinschaft; Comm. for the
Investigation of Health Hazards of Chem. Compounds in the Work Area. –
Weinheim ; New York ; Chichester ; Brisbane ; Singapore ; Toronto : Wiley-VCH
   Erscheint unregelmäßig. – Aufnahme nach Vol. 1 (1991)
Vol. 1 (1991) –

Composition: ProSatz Unger, D-69469 Weinheim.
Printing: betz-druck gmbh, D-64291 Darmstadt

Printed in the Federal Republic of Germany.

Preface

The Working Group "Analytical Chemistry" of the Commission for the Investigation of Health Hazards of Chemical Compounds in the Work Area is continuing its efforts to elaborate, validate and introduce suitable methods for determining hazardous chemicals in workplace air. At present, more than 120 methods have been published in the German edition of Analyses of Hazardous Substances in Air. Since Volumes 1 and 2 of the English edition aroused considerable interest, the Working Group has provided another series of 14 standardised methods in this volume. Volumes 4 and 5 are already in preparation. I wish to acknowledge the careful work of the group "Analytical Chemistry" and the contribution made by the authors of these methods, as well as the accurate work of the translators and editors of this volume. Again, I hope that this publication will be beneficial to the protection of the health and will improve the working conditions of those exposed.

H. Greim
Chairman of the Commission for the
Investigation of Health Hazards of
Chemical Compounds in the Work Area

Foreword

The protection of workers against risks from chemical agents is particularly important because of the many effects on the individual which may be caused by exposure to harmful chemical agents at the place of work. In 1995, the European Commission decided to set up a Scientific Committee to give advice on the setting of Occupational Exposure Limits (OEL) based on scientific and technical data. By applying a well-defined guidance note on procedures to set limit values, recommendations for more than 80 OELs have been made to the Commission. If the aim of the limit values to protect workers' health is to be realised, then it is essential that measurements of exposure should be reliable. Persons who carry out measurements must possess the necessary expertise and facilities. Furthermore, the measuring procedure used, including limit of detection, sensitivity, precision and accuracy, must be appropriate to the agent to be measured, its limit value and the workplace atmosphere.

Research and development of such procedures suitable for routine use, are the objectives of the Working Group "Analytical Chemistry" of the Commission for the Investigation of Health Hazards of Chemical Compounds in the Work Area. In response to the worldwide demand for reliable chemical methods for air analysis, the Working Subgroup "Analyses of Hazardous Substances in Air of Work Area" has chosen methods for publication, whose analytical reliability and reproducibility have been tested and confirmed by at least one examiner. The description of each method includes an evaluation of the method, a brief listing of the reliability criteria and general information on the chemical compound to be tested, i.e., its industrial importance, toxicity, and its limit value at the workplace as far as it is known. This is followed by a detailed description of the preparatory and analytical steps, discussion of the reliability and a reference list.

Volume 3 comprises further 14 analytical methods for determination of hazaardous substances at the workplace. Besides the description of analytical methods, this volume contains a chapter on procedures for the preparation of calibration gas mixtures. The provision of accurate, reproducible and controlable standards is an essential requirement for the determination of hazardous gases and vapours in ambient air. The general principles and techniques for preparation of gas mixtures are summarised in this chapter.

We would like to thank the members and guests of the Working Subgroup without whose voluntary services this collection of methods would not have been possible. We thank the Deutsche Forschungsgemeinschaft for financial and organisational help in the development of this project. Our thanks go also to our publisher Dr. Eva E. Wille of WILEY-VCH with whom we have enjoyed long-standing and efficient collaboration. We also wish to thank Dr. Brown, Mrs. Handwerker-Sharman, Dr. Ohrbach, and Dr. Wild for translation.

J. Angerer
Chairman of the Working Group "Analytical Chemistry" of the Commission for the Investigation of Health Hazards of Chemical Compounds in the Work Area

A. Kettrup
Chairman of the Working Subgroup "Analyses of Hazardous Substances in Air of Work Areas"

Contents

Contents

Working Group "Analytical Chemistry" of the Commission of the Deutsche Forschungsgemeinschaft for the Investigation of Health Hazards of Chemical Compounds in the Work Area

Organization

The Working Group "Analytical Chemistry" was established in 1969. Under the chairmanship of Prof. Dr. J. Angerer at present it includes two Working Subgroups:

"Air Analyses"
(Leader: Prof. Dr. A. Kettrup)

"Analyses of Hazardous Substances in Biological Materials"
(Leaders: Prof. Dr. J. Angerer and Chem.-Ing. K. H. Schaller)

The participants, who have been invited to collaborate on a Working Subgroup by the leaders, are experts in the field of technical and medical protection against chemical hazards at the workplace.
A list of the members and guests of "Analyses of Hazardous Substances in Air" is given at the end of this volume.

Objectives and operational procedure

The two analytical subgroups are charged with the task of preparing methods for the determination of hazardous industrial materials in the air of the workplace or to determine these hazardous materials or their metabolic products in biological specimens from the persons working there. Within the framework of the existing laws and regulations, these analytical methods are useful for ambient monitoring at the workplace and biological monitoring of the exposed persons.
In addition to working out the analytical procedure, these subgroups are concerned with the problems of the preanalytical phase (specimen collection, storage, transport), the statistical quality control, as well as the interpretation of the results.

Development, examination, release, and quality of the analytical methods

In its selection of suitable analytical methods, the Working Group is guided mainly by the relevant scientific literature and the expertise of the members and guests of the Working Subgroup. If appropriate analytical methods are not available they are worked out within the Working Group. The leader designates an author, who assumes the task of developing and formulating a method proposal. The proposal is examined experimentally by at least one other member of the project, who then submits a written report of the results of the examination. As a matter of principle the examination must encompass all phases of the proposed analytical procedure. The examined method is then laid before the members of the subgroups for consideration. After hearing the judgement of the author and the examiner they can approve the method. The method can then be released for publication after a final meeting of the leader of the Working Group "Analytical Chemistry" with the subgroup leaders, authors, and examiners of the method.

Under special circumstances an examined method can be released for publication by the leader of the Working Group after consultation with the subgroup leaders.

Only methods for which criteria of analytical reliability can be explicitly assigned are released for publication. The values for inaccuracy, imprecision, detection limits, sensitivity, and specificity must fulfill the requirements of statistical quality control as well as the specific standards set by occupational health. The above procedure is meant to guarantee that only reliably functioning methods are published, which are not only reproducible within the framework of the given reliability criteria in different laboratories but also can be monitored over the course of time.

In the selection and development of a method for determining a particular substance the Working Group has given the analytical reliability of the method precedence over aspects of simplicity and economy.

Publications of the working group

Methods released by the Working Group are published in the Federal Republic of Germany by the Deutsche Forschungsgemeinschaft as a loose-leaf collection entitled "Analytische Methoden zur Prüfung gesundheitsschädlicher Arbeitsstoffe" (WILEY-VCH Verlag, Weinheim, FRG).

The collection at present consists of two volumes:

Volume I "Luftanalysen"

Volume II "Analysen in biologischem Material".

These methods are also to be published in an English edition. Volume 1 to 5 of "Analyses of Hazardous Substances in Biological Materials" have already been published. The work at hand represents the third English issue of "Analyses of Hazardous Substances in Air".

Withdrawal of methods

An analytical method that is made obsolete by new developments or discoveries in the fields of instrumental analysis or occupational health and toxicology can be replaced by a more efficient method. After consultation with the membership of the relevant project and with the consent of the leader of the Working Group, the subgroup leader is empowered to withdraw the old method.

Preliminary Remarks

Procedures for the preparation of calibration gas mixtures

Contents

Procedures for the preparation of calibration gas mixtures

1 Introduction

The term calibration gas mixture is defined by the VDI guideline 3490, which is published as a series of loose-leaves, as follows: in most cases a calibration gas mixture is a compressed mixture of gases which is generally made of a complementary gas and one or more constituents. The complementary gas is a pure gas or a mixture of gases and is commonly the major component of the calibration gas mixture, diluting the constituents which serve for calibration. Its type and purity must be appropriate for the application. The constituent is a gaseous or vaporous component of a calibration gas mixture of a known quality and quantity. It is used directly for testing and calibration [1].

According to this definition, a calibration gas mixture is a gas mixture which contains at least one constituent of exactly known concentration which serves for calibration.

Calibration gas mixtures are required for many applications and different kinds of problems. They are often used for the calibration of analytical instruments like process analysers for the determination of SO_2 or NO_x, gas chromatographs or gas phase infrared spectrometers. Moreover, calibration gas mixtures are applied in model experiments and processes in the gas phase (e.g. adsorption). Gas mixtures of known composition are also used in the toxicology e.g. for inhalation studies.

In air analysis, calibration gas mixtures are not only needed for calibration but also to check sampling devices and sampling methods for the measurement of hazardous substances. The compositions and concentrations of the calibration gas mixtures may vary widely. Relatively high concentrations are necessary in the field of workplace measurements whereas low concentrations are of interest in the monitoring of immissions.

A survey of the various methods for the preparation of calibration gas mixtures is shown in Fig. 1.

The individual methods and equipment for calibration gas mixture generation are described in detail in the book "Controlled Test Atmospheres" by G. O. Nelson [3]. A large number of review articles [2, 4, 5], monographs (e.g. [6] deals with permeation), guidelines [1, 7–28] as well as proceedings of symposia and workbooks deal with the different methods comprehensively and with various priorities (e.g. commercially available calibration gas mixture generators and equipment [5]). Often the information is widespread in the literature and only to be found in connection with a distinct measuring problem.

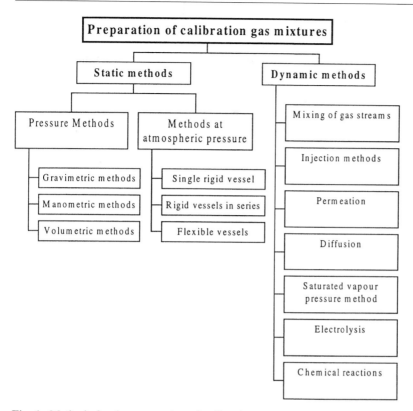

Fig. 1. Methods for the preparation of calibration gas mixtures (according to [2]).

2 General principles

2.1 Complementary gas requirements

The complementary gas and its constituents have to meet particular requirements. One of the most important requirements is that of adequate purity. Impurities can cause errors in the calibration or in the interpretation of the measuring results, but not necessarily in every case. For example, a calibration gas mixture for propane which is contaminated with hexane would lead to an additional peak but it would not interfere with the calibration of propane. An instrument for measuring the total hydrocarbon content calibrated with the same calibration gas mixture would always indicate too low. Especially in the generation of calibration gas mixtures of low concentration, the purity of the complementary gas is important. It is possible that argon 5.0 (argon of a purity of 99.999%), for example, contains up to 10 mL/m^3 impurities. Often purification of the complementary gas over activated carbon is necessary. Depending on type and purpose of the calibration gas mixture to be generated, drying of the calibration gas mixture may be useful. Besides air, nitrogen and argon are often used as complementary gas. These gases are available in high purity. As the calibration gas mixtures generated this way contain lower quantities oxygen or moisture, they are often more stable and can be stored for a longer period. However, if possible, the composition of the calibration gas mixture should correspond with the real sample to avoid systematic errors (matrix effect). For example, it is well known [29] that the accurate flame photometric calibration of sulfur dioxide requires the same CO_2 concentration in the calibration gas mixture as in the airborne sample. Depending on the purpose of the calibration gas mixture it may be necessary to be humidified to a known content.

2.2 Constituents requirements

The purity requirements for the constituents are similar to those for the complementary gas. The impurity of the constituent mostly influences the concentration of the calibration gas mixture in per cent. However, the influence may be greater with different methods (e.g. saturated vapour pressure method, permeation method).

In the selection of the concentration of the constituent a possible concentration limitation has to be considered for physical or safety technical reasons. For example, condensation may occur under pressure at high concentrations. In case of constituents which may react with the complementary gas in a particular concentration range (explosion limit, ignition limit) it has to be considered that an increase in pressure shifts these limits to lower concentrations. At 8 MPa the maximum permitted concentration of propane in air is 1.0% by volume whereas at 15 MPa the maximum concentration is only 0.5% by volume [7].

2.3 Apparatus requirements

Errors in the generation of calibration gas mixtures may be caused by the apparatus or by auxiliary equipment used. For this reason the apparatus, its components and tubing must fulfill particular requirements:

- Tightness of the complete apparatus for the generation of calibration gas mixtures.
- Inertness of the materials used to the chemicals.
- Short tubing, if possible, all connections should have minimal dead space volumes.
- Reducing of the wall adsorption of the apparatus (especially important in static methods) e.g. by silanization of glass, electro-polishing of metal or coating of the internal walls.
- Purity of the apparatus; an impurity of the calibration gas mixture e.g. by stopcock grease or by constituents used before have to be avoided.
- The time required for the system to reach the equilibrium has to be awaited before each measurement or each calibration.

- Exact and precise determination of the parameters such as volumes, pressure and flow rate which are necessary for the calculation of the calibration gas mixture concentration; the flowmeters should be checked regulary e. g. with a soap bubblemeter.
- Temperature constancy is important in many methods, it is sometimes necessary to keep the temperature of a constituent constant at ± 0.1 K.
- If the calibration gas mixture has to be stored for later use and the shelf-life has not yet been determined, the composition of the stored calibration gas mixture has to be checked at regular intervals (shelf-life may be affected by vessel material, moisture and atmospheric oxygen).

2.4 Calibration gas mixture requirements

A calibration gas mixture should meet the following requirements [4]:

- The calibration gas mixture should be stable, i. e. a defined composition of the calibration gas mixture should be constant for long periods of time.
- It should be available in amounts sufficient to carry out all necessary investigations.
- In addition, the accuracy with which the calibration gas mixture composition is determined should be greater by a factor of 2.5–3 than the accuracy of the system being checked or calibrated.
- Impurities should not influence the measuring result.
- Fluctuations in the composition of the calibration gas mixture should not be greater than 2–5%.
- Preparation of calibration gas mixtures should make use only of fundamental parameters such as weight, temperature and pressure.
- The shelf-life of the calibration gas mixture should be known.
- All possible sources of error should be defined exactly and their sizes should be known.

Although the exact requirements to be met by calibration gas mixtures depend on the particular application, their composition must always be known with sufficient accuracy. In general, during generation of calibration gas mixtures, the quantity of the constituent is measured as weight or volume, the quantity of complementary gas is measured as volume or flow rate and the calibration gas mixture concentration can generally be calculated from these parameters. In many cases the ideal gas law $pV = nRT$ is applied. Deviations from the ideal gas behaviour make it necessary to make corrections for internal pressure and covolume, especially at high pressures and low temperatures. Sometimes they can be calculated [30].

To achieve low calibration gas mixture concentrations it is often appropriate to generate a calibration gas mixture at a higher concentration and then to dilute this gas with complementary gas. Homogeneity of the calibration gas mixture requires thorough mixing (see the remarks in the individual sections).

3 Static methods

In general, the various methods for generation of calibration gas mixture may be divided into static or dynamic methods (cf. Fig. 1).

In the static methods, a defined quantity of a constituent is transferred into a vessel of a known volume containing the complementary gas. After evaporation and mixing with the complementary gas, a homogeneous calibration gas mixture is obtained. The concentration can easily be calculated. At ambient pressure static methods are suitable for generating small volumes of calibration gas mixture. They do not require complex instrumentation and can be used for relatively high concentrations of calibration gases. However, at very low concentrations in rigid vessels the shelf-life can be very limited. Static methods with overpressure are often used for commercial calibration gas mixtures. The great disadvantage of the static methods are adsorption and condensation effects on the walls of the used vessels which as a result of the limited volume may lead (especially at low concentrations) to significant errors due to the removal of the analyte in the calibration gas mixture. In stored calibration gas mixtures, concentrations should be checked regularly because many different kinds of losses may occur (wall adsorption, reactions, diffusion).

3.1 Methods at atmospheric pressure

As mentioned above these methods are suitable for small volumes of calibration gas mixtures and higher concentrations (preferably > 1%). They do not require complex apparatus, are generally quick to carry out and sufficient for many applications.

3.1.1 Single rigid vessels

One of the simplest methods of generating calibration gas mixtures makes use of rigid vessels of glass, plastic or metal. Glas bottles, gas sampling bulbs ("gas mice"), gastight glass syringes, polyethylene bottles, plastic containers, metal cylinders, flasks or, for example, spaces with metal linings are used (cf. Fig. 2 a and b). The exact volume of the rigid vessel must be known. It is determined experimentally by volume determination, or for large containers, calculated from the dimensions in both cases temperatures and pressure must be taken into consideration. If necessary, the vessel is cleaned and then flushed several times with the complementary gas. At a constant temperature it is then filled with the complementary gas. The resulting calibration gas mixture in the vessel has to be at atmospheric pressure (neither overpressure nor underpressure). The constituent is added as liquid or gas. A measured quantity (volume or weight of a gas or liquid) of a pure substance or mixture of known composition is transferred into the vessel. (The constituent may also be injected into an evacuated vessel and then to bring the pressure to ambient with complementary gas. The substances are injected through a septum with a gastight syringe.). High boiling liquids can be vaporized by heating. A stirrer or a fan improve mixing; for smaller vessels a magnetic stirrer may be used or the constituents

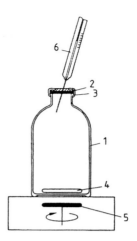

Fig. 2 a. "Gas mouse" (gas sampling bulb) with teflon stopcock and sidearm port with exchangeable septum.

Fig. 2 b. Diagram of a rigid vessel for the generation of calibration gas mixtures. 1 100 mL crimp-top vial, 2 PTFE coated butyl septum, 3 crimp cap, 4 magnetic stirrer, 5 stirring device, 6 microliter syringe for adding constituents (according to [31]).

mixed with glass beads. After complete evaporization of the constituents and mixing, the calibration gas mixture can be taken from the vessel (e.g., through a septum).

Static calibration gas mixtures at ambient pressure do not have long shelf-lives. A study of the shelf-life of various calibration gas mixtures in teflon vessels (in the ppb range) has been published [32]. Calibration gas mixtures in glass vessels at concentration below 1 mL/m^3 should only be prepared immediately before use [19].

If a single rigid vessel is used to prepare the calibration gas mixture, only a very limited fraction of the volume is usually available. In most cases only about 10% of the volume can be taken from such a vessel [3] because of excessive underpressure or dilution with air. Adsorption to walls and connections is a disadvantage of this method, causing large errors especially at low concentrations (ppb, ppt range). Silanized glass has shown to be the vessel material causing the least wall adsorption and chromium-molybdenum steel the most [33].

3.1.2 Rigid vessels in series

The connection in series of several rigid vessels of the same type containing the same calibration gas mixture is a method of enhancing the available calibration gas mixture volume. In this arrangement, instead of air a stream of calibration gas mixture from the adjacent vessel flows in to replace the removed calibration gas mixture. Provided that no significant adsorption at the connection fittings occurs, the connection of 5 rigid vessels

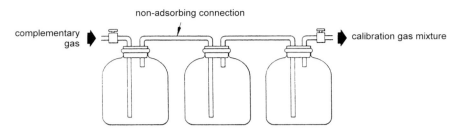

Fig. 3. Diagram of rigid vessels connected in series [3].

in series (cf. Fig. 3), for example, permits the removal of the 2.5-fold of the volume of one vessel before a dilution effect may be observed which is comparable with that caused by the removal of 10% of the volume of one individual rigid vessel. The concentration in the system after the removal of a given volume calculated as follows [3]:

$$C_n = C_0 [1 + 1/1! \cdot (V_w/V_b) + 1/2! \cdot (V_w/V_b)^2 + \ldots + 1/(n-1)! \cdot (V_w/V_b)^{n-1}] \, e^{-V_w/V_b}$$

Legend:

C_n Concentration in the vessel n after the removal of the volume V_w
C_0 Initial concentration
V_w Volume which is removed
V_b Volume of one vessel
n Number of the vessels

3.1.3 Exponential dilution

Exponential dilution is intermediate between the dynamic and static generation of calibration gas mixture. Applying this method, first a static calibration gas mixture with a defined initial concentration c_0 is generated in a rigid vessel. The vessel is fitted with an additional port for the complementary gas and a stirrer (e.g. magnetic stirrer) (cf. Fig. 4) for mixing. If the calibration gas mixture is removed continuously from the vessel the complementary gas is allowed to stream into the vessel and the original mixture is diluted exponentially. Provided mixing is thorough, the gas concentration c in the vessel my be calculated at any time as follows [34]:

$$c = c_0 \exp(-FT/V)$$

Legend:

F Flow rate through the vessel
T Time of the removal
V Volume of the vessel
c_0 Initial concentration

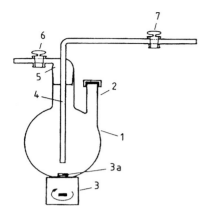

Fig. 4. Apparatus used for the dilution method (according to [31]). 1 Two-necked glass flask, 2 neck septum for adding the constituent, 3/3 a stirring unit with magnetic stirrer, 4 inlet for the complementary gas, 5 sampling head, 6/7 stopcocks.

At time T = 0 the concentration c is equal to the initial concentration c_0 and then decreases exponentially to zero.

Constant concentrations cannot be prepared with exponential dilution systems. If calibration gas mixtures are used in experiments where memory effects are to be expected, errors may result from the very rapid change of the concentration at the beginning of the experiment.

The concentration profile obtained in the exponential dilution method can be modified by connecting two vessels in series (cf. Fig. 5 a and b). In this case a vessel containing only complementary gas is connected to the calibration gas mixture outlet of the vessel with the exponential dilution. The calibration gas mixture is taken from this vessel con-

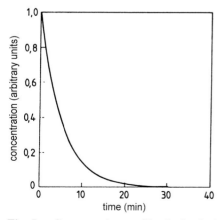

Fig. 5 a. Concentration profile obtained with a single exponential dilution [34].

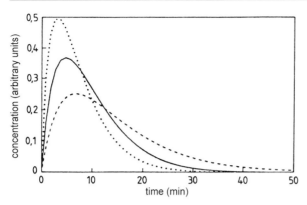

Fig. 5 b. Theoretical concentration profile obtained with exponential dilution through two rigid vessels in series: continuous line: $V_1 = V_2$, dashed line: $V_2 = 2 V_1$; dotted line $V_1 = 2V_2$ [34].

taining the complementary gas. On the condition that the volumes in both vessels are identical, the available concentration c is calculated according to the equation:

$$c = kc_0 T \exp(-FT/V)$$

whereas k is an integration constant. The calibration gas mixture concentration starts at zero, increases to a maximum and then decreases exponentially to zero [34].
It must be emphasized that concentrations which are constant for long periods cannot be prepared with the exponential dilution method.

3.1.4 Flexible vessels

Flexible gastight plastic bags of different materials, often lined with PTFE, aluminium, polyester or PVC, are also used as vessels (cf. Fig. 6). For the generation of a calibration gas mixture they are flushed with the complementary gas with which they are then

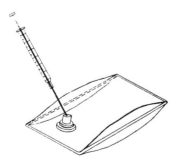

Fig. 6. Flexible sample bag.

filled either completely or to a known volume. Any overpressure which develops during the filling procedure can be equalized through a septum pierced with a cannula. As with the "gas mice", the constituent is added through the septum with a syringe. In case of smaller bags (up to a volume of about 5 L) and small differences in the densities of the complementary gas and the constituent, storage of the filled bag for half an hour is sufficient to homogenize the mixture [16]. If the bags are only partially filled homogenization can be done by kneading. When the densities of the complementary gas and the constituents are very different or if the total volume is larger, the bag should be placed on end and its lower half heated in a stream of hot air. The temperature has to be selected according to the thermal stability of the plastic bag and e.g. it should not exceed 60 °C for plastics containing polyethylene [16].

Samples may be removed from flexible vessels without diluting the calibration gas mixture. Therefore the whole volume can be used. Besides the wall adsorption of the flexible thin-walled vessels constituental substance losses are possible by diffusion through the vessel wall. Both effects are increased when calibration gas mixtures are stored. The material of the bag and its lining have to be selected carefully depending on the kind and concentration of the constituents.

The vessels must be very clean inside. Memory effects are observed more often than with glass vessels; they can lead to application errors.

3.2 Pressure methods

These methods are used almost exclusively applied by commercial suppliers of calibration gas mixtures. The gas cylinders are seamless drawn or pressed by extrusion, with volumes normally between 10 and 200 L. In most cases the material consists of an aluminium alloy or tempered steel [35]. Depending on the purpose the bottles are pretreated in various ways. Mostly they are intensively heated and evacuated before filling. Temperature fluctuations and especially the compressibility of the gases which lead to deviations from the ideal gas law have to be taken into account when filling the cylinders. During generation of the calibration gas mixture, the individual ignition limits have to be considered and condensation (e.g. at low storage temperatures) avoided. The cylinders can be filled with compressed gas in three different ways which are described in detail in the following sections. In order to achieve a homogeneous mixing of the calibration gas mixture, the cylinders are turned mechanically. The mixing can be accelerated by slightly warming the vessel at one side. Stable mixtures can be prepared if the substances used are chemically stable and if they do not react with the other constituents of the mixture or with the vessel wall. The internal surface of the vessels and their valves are pretreated to prevent reactions with the constituents of the gas mixture and to reduce adsorption effects. Therefore, the internal surface of the vessel is treated mechanically and chemically, cleaned, flushed with inert gas and finally evacuated [36].

Users should be aware of the minimum and the maximum storage temperature, the maximum shelf-life and the minimum pressure of a compressed gas as well as its composition [1]. A study of the stability of various mostly organic compounds compressed in gas cylinders has been published [37].

3.2.1 Gravimetric methods

With the gravimetric method, the individual constituents are filled consecutively into a gas cylinder and each weight increase is determined exactly with a special balance. The weighing may be carried out in a vacuum or at ambient pressure. In the latter case, the gas cylinders must be allowed to reach ambient temperature. In addition, the measured weight has to be corrected for any changes in buoyancy of the vessel. A prerequisite for the gravimetric measurement is the availability of balances which can indicate large total weights with high accuracy and sensitivity (e. g. 30 kg $\pm$ 0.1 mg exactly) [4]. Since too small quantities of gases cannot be weighed exactly enough a lower limit X_0 is determined for the concentration of each constituent. This value depends on the sensitivity and the maximum load of the balance as well as on the inaccuracy in the concentration of the constituent which can be tolerated. If the desired concentration is lower than this limit X_0, a mixture of a higher concentration is prepared at a higher concentration and then diluted with a known quantity of complementary gas [9].

Among others the gravimetric method is applied for the generation of commercially available primary calibration gas mixtures [38] because this method normally offers the highest accuracy of all overpressure methods. The generation of the calibration gas mixtures and the pretreatment of the gas cylinders are very time-consuming processes [39]. In addition, the calibration gas mixture has to be tested very extensively. A standard reference material SRM 1804 is available from the EPA. It contains 18 volatile mainly halogenated organic compounds in pure nitrogen [40]; it is intended for testing ambient air measuring systems and required years of research work for its development. It is documented, for example, that 15 of the calibration gas mixture constituents remained homogeneous and stable for two years [40]. The individual substance concentrations in this standard are very low i. e. in the range of 5 nmole/mole with a total uncertainty between 2 and 6%.

3.2.2 Manometric methods

In the manometric method, the pressure changes are measured after the addition of each individual constituent and the complementary gas. The sequence in which the constituents are added depends on the pressure of the gas supply (filling pressure, vapour pressure); generally the constituent available at the lowest pressure is added first (cf. Fig. 7). On the other hand, it is advisable to add first the constituent whose behaviour deviates most from the ideal because the effect of such deviations generally increases with increasing pressure [17].

Above all, in the manometric method too, the different compressibilitiy of the gases used has to be taken into considerations as well as temperature changes and deviations from ideal behaviour. One method uses the Benedikt-Webb-Rubin equation, an empirical equation which requires a very large number of pVT data. For this reason, it can only be applied for substances for which data is available [41].

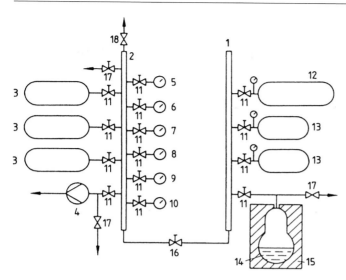

Fig. 7. Diagram of a manometric filling system for the generation of calibration gas mixtures [17]. 1 Supply manifold, 2 filling manifold, 3 gas cylinders for calibration gas mixture, 4 vacuum pump, 5–10 manometers, 11 shut-off valve, 12 gas cylinder for complementary gas, 13 gas cylinders for the constituents, 14 container for highly volatile liquified gas, 15 cooling device (if necessary), 16 control and shut-off (stop) valve, 17 safety valves, 18 flushing and bleed valve.

3.2.3 Volumetric methods

The volumetric method includes the dosing of exact volume parts as well as the mixing of volume streams.
In the first method for the volumetric generation of a calibration gas mixture, a gas cylinder of known volume is filled with a gaseous constituent at ambient pressure and a given temperature. Then the complementary gas is added as dilution gas, the weight of which normally determined gravimetrically.
However, the dynamic volumetric methods are used more frequently (cf. Sect. 4.1). A gas mixture prepared dynamically can then be compressed and filled into gas cylinders. An advantage of this method is that a series of gas cylinders may be filled with the same mixture [41].

4 Dynamic methods

The dynamic methods for the generation of calibration gas mixtures have gained great importance despite the great technical complexity of the apparatus required. In contrast to the static methods, in the dynamic methods the carrier gas and the constituent are mixed in a dynamic flow which yields a continuous supply of the calibration gas mixture. An advantage of the dynamic methods is that the practically unlimited volume of calibration gas mixture sets no limits on the time available for calibration and further investigations. Although dynamic methods require complex apparatus, they often lead to more accurate calibration gas mixture concentrations. Another advantage is that it is relatively simple to change the concentration. In addition, short equilibration times facilitate adaptation of the method to the current task. Wall adsorption also occurs in the dynamic methods, but after saturation it is practically insignificant. On principle, the apparatus has to run in for a sufficient period before analysis can begin. The generation of a continuous gas stream is appropriate in particular for reactive calibration gas mixtures with which storage problems usually occur in static systems.

4.1 Mixing of gas streams

The mixing of gas streams is an important method for the dynamic generation of calibration gas mixtures; it is also combined with other methods such as permeation, diffusion or saturated vapour pressure methods because it is suitable for dilution of calibration gases.

The gaseous constituent is mixed to a constant stream of complementary gas, both flowing at constant rates, are combined so that a defined mixture is obtained. For homogenization, the mixture is generally conducted through a mixing chamber which can be constructed in various ways. Then a continous supply of a homogeneous calibration gas mixture at ambient pressure is available.

The quality of the resulting calibration gas mixture is determined by the accuracy and constancy of the indvidual flow rates. Normally the flow control is relatively complex and can be a principal source of error. At small flow rates often mass flow controllers are used. It is also possible to use a computer for the monitoring and electronic adjustment of the flow control, this can yield good accuracy [42]. On principle, the critical parameter is not the volume flow rate but the mass flow rate which is determined from the density of the flowing gas. In addition, calibrated flowmeters or capillaries are often used for the adjustment of the flow rates. Flow control may also be achieved with diaphragms, critical orifices and pumps a. o.

The arrangement for mixing gases shown in Fig. 8 permits several gaseous constituents to be added to the complementary gas at constant pressure via a mixing line with orifices. The orifice diameters are in the range of $5-1000$ µm.

In systems with gas mixing pumps (piston dosing pumps) two pumping pistons are driven over several gearwheels by a synchronous motor. At the same pressure the gases

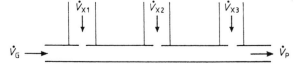

Fig. 8. The addition of several other constituents to the stream of complementary gas via a mixing line equipped with orifices [21]. V_G stream of complementary gas, $V_{X1}-V_{X3}$ streams of the other gaseous constituents $(1-3)$, V_P stream of calibration gas mixture.

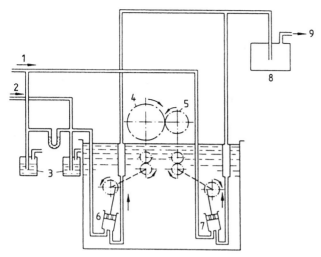

Fig. 9. Gas mixing pump [11]. 1 Inlet for gas B, 2 inlet for gas A, 3 pressure compensation, 4 driving wheel for cylinder a, 5 driving wheel for cylinder b, 6 cylinder a, 7 cylinder b, 8 mixing chamber, 9 calibration gas mixture outlet.

are drawn by the pumping pistons through a mixing chamber to the user (cf. Fig. 9). The number of strokes per unit of time is kept constant so that the concentration of the calibration gas mixture which is generated remains constant. Exchangeable gearwheels and gear units for the adjustement of the mixing ratio permit a specific change of the flow rates and generation of different concentrations.

Constituent concentrations lower than 1% by volume can be achieved by connecting several mixing pumps in series or by two-stage or multi-stage mixing of the gas streams. Such arrangements also allows the generation of calibration gas mixtures in the ppt range [43]. The concentration of the resulting calibration gas mixture is calculated from the ratio of the flow rate of the constituent and the total flow rate. Temperature and pressure have to be taken into consideration.

4.2 Injection methods

In the injection methods the gaseous or liquid constituents are directly injected into the complementary gas with a suitable injection devices. Continuous and periodic processes must be distinguished. For example, periodic injection of constituents can be carried out via injection stopcocks or injection loops. The main disadvantage of such pulsating systems is the short-term periodic fluctuation in concentration which makes large mixing chambers necessary. The volume required depends on the total gas flow rate and has to be a multiple of the gas volume which flows into the chamber between two additions of analyte [8]. Continuous injection can be performed, for example, with motor-driven syringes or pistons. Motor driven injection devices may be used over wide concentration ranges; the concentrations can be varied rapidly by changing the feed rate. As in the static methods, when liquids are injected, complete evaporation of the injected substance must be ensured.

4.2.1 Injection stopcocks and injection loops

Injection stopcocks and injection loops are used in the periodic injection methods. An injection stopcock is shaped like a four-way stopcock. The tap plug is turned by an impulse-controlled motor. The stopcock bore, with its defined internal volume, joins the opposite stopcok fittings (cf. Fig. 10).

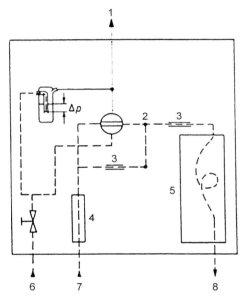

Fig. 10. Mixing station with injection stopcock (see [44]). 1 Waste, 2 injection stopcock, 3 capillary, 4 flowmeter, 5 mixing chamber, 6 constituent, 7 complementary gas, 8 calibration gas mixture.

First the complementary gas streams into the mixing chamber through the valve or a capillary by-pass. The analyte flows to waste through a wash bottle which serves for pressure regulation. When the connection is opened by turning the stopcock clockwise, the constituent streams through the stopcock and flushes it. When the stopcock is turned again, a defined volume of the constituent is carried with it. After a tap rotation of 90°, the stopcock bore is flushed with the complementary gas which carries the defined volume of constituent into the mixing chamber.

If injection loops are used instead of injection stopcocks, the system is controlled via a camshaft. Four valves are installed at the connections and are switched in pairs so that the injection device is flushed alternately with the constituent and the complementary gas. The concentration of the resulting calibration gas mixture depends on the volume of the injection stopcock or injection loop, on the number of the injections per unit of time and on the flow rate of the complementary gas. It may be difficult to determine the volume of the stopcock. For this purpose dynamic methods are often chosen.

4.2.2 Piston injection device

A gastight glass piston permits the injection of a gaseous constituent through a capillary into a continuous stream of complementary gas. In most cases the piston injection device is controlled by mechanical gears and injects the constituent continuously (cf. Fig. 11). A mechanically driven syringe can also be used for the injection of liquids.

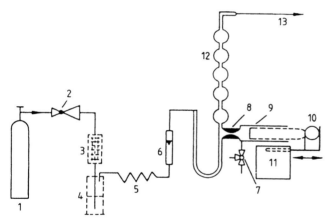

Fig. 11. Piston injection device [13]. 1 Gas cylinder with complementary gas supply, 2 pressure control, 3 cleaning vessel (if necessary), 4 humidifier (if necessary), 5 spiral tube for temperature compensation, 6 flowmeter, 7 three-way stopcock, 8 capillary, 9 constituent, 10 piston, 11 piston drive, 12 mixing tube, 13 calibration gas mixture outlet.

The period for which such a system can deliver calibration gas mixture is limited by the volume of the syringe and the drive velocity; for example, with a syringe volume of 20 to 100 mL it may be limited to about 8 hours. When liquid substances with a boiling point below 40 °C are injected, bubbles may be formed at the tip of the injector. This problem may be solved by keeping the temperature constant with a cooling jacket [45]. For higher boiling liquids (above 120 °C), it is necessary to check that the substances evaporate completely. A plug of glass wool near the tip of the injection needle often increases the efficiency of evaporation.

A saturated gaseous constituent can be prepared, for example, by wetting the internal surface of the piston cylinder with the liquid constituent so that a saturated gas is formed within a short time. It is important that the liquid only wets the cylinder walls and that it is not injected [46]. Low concentrations can be prepared in a subsequent dilution step or by diluting the constituent with a solvent before injection. The solvent should not interfere with the later application [16].

4.2.3 Capillary injection device

In this method, the constituent is injected into the continuously flowing complementary gas through a capillary of known length and known internal diameter (cf. Fig. 12). The constituent to be injected is kept at the lower end of a vertical capillary at a pressure which is higher by a defined amount than that at the upper end of the capillary. Due to the pressure difference, the constituent streams through the capillary. At the upper end

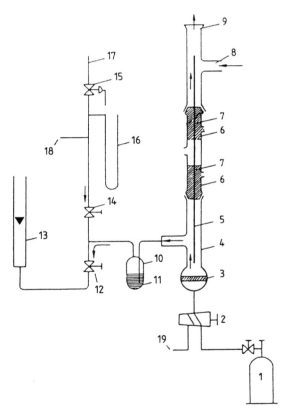

Fig. 12. Capillary injection device for gases [15]. 1 Gas cylinder with pressure reducing valve and needle valve for the constituent, 2 two-way stopcock, 3 gas filter (e. g. glass frit G4), 4 overflow tube, 5 calibrated injection capillary, 6 capillary tube with compensation opening, 7 seal, 8 complementary gas inlet, 9 mixing line with calibration gas mixture outlet, 10 bubblemeter, 11 barrier liquid, 12 throttle valve, 13 flowmeter, 14 stopcock, 15 pressure control, 16 manometer, 17 inert gas supply, 18 inert gas outlet for flushing, 19 inert gas inlet for flushing.

of the capillary the constituent is vaporized (if it is a liquid) and is mixed with the stream of complementary gas to make up the defined calibration gas mixture. The injected flow rate of the constituent may be calculated with the Hagen-Poiseuille equation with corrections for compressibility and shearing strain [47]. The following parameters have to be known: dimensions of the injection capillary, pressure difference between the ends of the capillary, viscosity and density of the injected gas or liquid constituent. Pure liquids, pure gases or even mixtures of defined composition may be used [15]. When liquids are used, the concentration may be determined by continuous or discontinuous measurement of the decrease in weight of the supply vessel during generation of the calibration gas mixture [48].

Errors may result from variations in temperature or a change in the diameter of the capillary due to residues deposited, e. g., from evaporating liquids.

4.3 Permeation

In the permeation methods for generation of calibration gas mixtures, the constituent is added continuously to the stream of complementary gas through an appropriate membrane. A pure substance or a mixture can be added; the substances can be gaseous, liquid or solid. Depending on its properties, the constituent or a suitable chemical precursor can be added as a solution, an addition compound, a molecular or other complex or an adsorbate [14]. The constituent is separated from the complementary gas by a membrane (cf. Fig. 13 a–c) of a polymeric material, e. g., PTFE, polyurethane, polyethylene or polypropylene [6]. The membrane can vary in diameter, width and thickness. For the generation of high concentration calibration gas mixtures, for example, PTFE tubing is used as the membrane because long tubes can provide a large surface area [49]. There

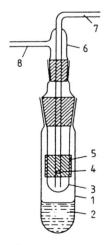

Fig. 13 a. Permeation finger with flat membrane [14]. 1 Supply vessel, 2 constituent supply, 3 finger, 4 orifice, 5 permeation membrane, 6 head-piece, 7 inlet tube for the complementary gas.

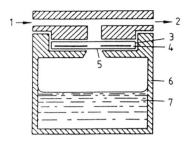

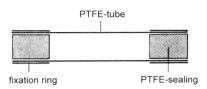

Fig. 13 b. Permeation vessel with flat membrane [14]. 1 Complementary gas inlet, 2 calibration gas mixture outlet, 3 permeation membrane, 4 metal orifice, 5 permeation area, 6 supply vessel, 7 constituent supply.

Fig. 13 c. Permeation tube.

are various types of permeation vessels. Some of them are refillable. There are also commercially available permeation tubes (e. g. Fig. 13 c) with defined permeation rates. After use for a limited period, they are disposed of. These tubes often contain liquified gas. Therefore their application is limited to gases with critical temperatures of 20–25 °C and moderate critical pressures [6].

A small number of molecules reach the complementary gas through the membrane by continuous permeation. The driving force in this diffusion process is the partial pressure gradient of the constituent between the internal and the external surfaces of the membrane. The quantity of substance transferred depends on the thickness and the area of the membrane as well as on the membrane material. It also depends on the kind of constituent and the concentration difference between the two sides of the membrane. The temperature dependence is very distinct and is described by the Arrhenius equation. If the concentration of a calibration gas mixture is to be stable within 1%, the temperature must be constant within ≤ 0.1 K [4].

The permeation rate itself cannot be calculated. On the one hand, the permeation is reduced by back diffusion and, on the other hand, the calculation would require ideal membranes and complete thermodynamic equilibrium as well as ideal gas behaviour [14]. The permeation rate may be determined experimentally in a number of ways: volumetrically, manometrically, colorimetrically, coulometrically or gravimetrically [4]. In most cases the absolute gravimetric method is applied, that is, the amount of substance permeated is determined at intervals by regular repeated weighing of the supply of constituent. With permeation systems, after the system-dependent equilibration period, which may be shortened by previous swelling of the membrane [6], constant calibration gas mixtures can be generated with relatively little maintenance for long periods of time. The production period is limited mainly by the supply of the constituent. Sources of error include: the blocking of pores with a condensed or polymerized constituent, contamination of the permeable membrane during installation or high humidity during storage of the membrane or during generation of the calibration gas mixture [2]. Dry air is generally used to generate calibration gas mixtures with this method because the swelling of the membrane varies with the moisture content.

Above all, low concentrations of calibration gas mixtures can be generated with the permeation method. It is not necessary to use high complementary gas flow rates. It is an advantage that it is possible to generate reactive gases such as SO_2, NO_2, H_2S [50, 51] or formaldehyde [52].

4.4 Diffusion

In the generation of calibration gas mixtures by diffusion, the gaseous, vaporized phase over a liquid constituent diffuses through a capillary into the stream of complementary gas (cf. Fig. 14a). Diffusion tubes of various diameters are commercially available [45] or can be made of pieces of uncoated gas chromatographic capillary columns [53]. If mixtures of several constituents are to be generated, a separate diffusion cell has to be used for each constituent. Mixtures of liquids should not be used because the individual substances evaporate at different rates.

As with the permeation methods, low and constant concentrations of calibration gas mixtures can be generated for long periods [54]. A change of concentration can be achieved by the use of different lengths of diffusion tube or different diameters. The flow rate of the complementary gas and the temperature can also be varied. The relatively long time required for equilibration after a temperature change must be allowed for. It is often recommended that the temperature be kept as constant as possible (as for permeation: within 0.1–0.2 °C). If the complementary gas flow rate is varied, the diffusion rate may increase with increasing flow rate [55]. The amount of diffused substance is determined experimentally (gravimetrically) or calculated if the diffusion coefficient is known. The equation for the calculation is derived in reference [56]. The diffusion coefficient can also be determined experimentally. For this purpose a diffusion vessel with a graduated calibrated capillary can be used; the liquid is in the capillary and the change of the liquid level with time is measured (over several days) (cf. Fig. 14b and c). Diffusion methods differ from permeation methods especially in that permeation rates are in the ng/min range and diffusion rates in the μg/min range. Moreover, calibration gas mixtures can be generated by diffusion only from solid and liquid substances, not from gases [57].

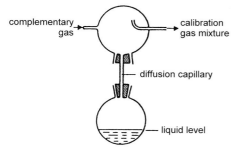

Fig. 14a. Diffusion apparatus according to McKelvey-Hoelscher.

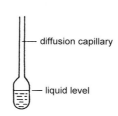

Fig. 14b. Diffusion vessel, simple.

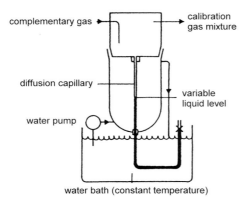

Fig. 14c. Diffusion vessel with calibrated capillary.

4.5 Saturated vapour pressure method

In these methods, the complementary gas is conducted along the surface of the constituent or through the constituent which, in most cases, is a readily vaporized liquid; the
constituent is enriched in the complementary gas. A variety of vessels can be used for
the enrichment step; for example, fritted bubblers are common.

The enrichment step in the saturated vapour pressure method may be carried out as a
one (cf. Fig. 15a [58]) [30] or two-stage (cf. Fig. 15b [59]) [18, 60] procedure. In the
former case, saturation is achieved very slowly; often the gas produced is only partially
saturated, its concentration depending on the residence time and thus on the flow rate.
In two-stage enrichment, two enriching vessels are connected in series. The temperature
in the second vessel is much lower (difference about 20 °C). The gas stream flowing

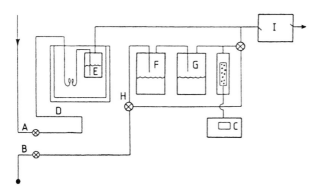

Fig. 15a. One-stage enrichment according to [58]. A Air supply to the constituent, B air for dilution, C moisture sensor, D constant temperature bath, E vessel with the constituent, F, G water
containers for humidification, H by-pass line, I mixing chamber.

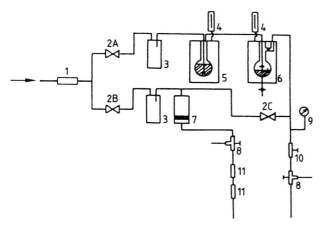

Fig 15 b. Two-stage enrichment [59]. 1 Molecular sieve, 2 A-C thermal flow regulators, 3 non-return valve, 4 thermometer, 5 container with liquid component, 6 cooling trap, 7 mixing chamber, 8 three-way stopcock, 9 manometer, 10 needle valve, 11 adsorption tubes.

from the first vessel is oversaturated at the low temperature in the second; the excess substance condenses or freezes out. The resulting calibration gas mixture is saturated, that is, the concentration is reproducible at the given temperature and pressure. Within certain limits, the concentration is independent of the flow rate. In a subsequent dilution step, the concentration of the calibration gas mixture can be decreased by adding a defined amount of complementary gas. The advantage of this method is that a constant and reproducible calibration gas mixture with a calculable composition can be generated. The concentration may be determined experimentally by gravimetric determination of the decrease in the supply of the constituent (one-stage method) or by calculation with the general gas equation, knowing the appropriate partial pressure (vapour pressure) of the constituent at the temperature of the cooling trap (taken from tables or diagrams [18, 61–64]). As in the permeation methods, a very constant constituent temperature is the condition for sufficient reproducibility and constancy of the calibration gas mixture. If this is ensured, the calibration gas mixture concentrations generated by this method remain constant for several days [65].

This procedure generates more concentrated calibration gas mixtures than do the diffusion and the permeation methods. Concentration changes may be brought about by variation of the constituent gas flow rate or temperature.

Saturated vapour pressure methods are often used to humidify calibration gas mixtures.

4.6 Electrolysis

Atmospheric gases such as CO_2, NO_2, N_2, SO_2 and O_3 can be generated in low concentrations by electrolysis. This requires that a sufficiently high current in the microampere or milliampere range flows between two inert electrodes (e. g., platinum) which are immersed in a liquid (cf. Fig. 16a and b). For example, hydrogen (cathodic) and oxygen (anodic) can be generated from acidic or basic aqueous solutions, carbon dioxide (CO_2) from aqueous oxalic acid and ozone from sulfuric acid. The generated gases can be conducted as constituents into the stream of complementary gas. The concentration can be varied by varying the current intensity and the flow rate of the complementary gas. As both anodic and cathodic gases are formed, the apparatus has to be constructed so that one of the gases can be disposed of, if it would otherwise interfere with the calibration gas mixture.

The concentration of the generated calibration gas mixture can be determined arithmetically. It is proportional to the mole volume of the generated gas and the applied electrolysis current and inversely proportional to the flow rate of the complementary gas, Faraday's constant and the number of electrons exchanged at the electrodes. Correct functioning of the process requires that the electrolyte is pure and present in excess. For some electrolytic processes, a correction factor has to be applied in the calculation of the concentration. The correction factor has to be determined with an independent method since it is affected, for example, by oxidation of the electrode (discoloration) and also by impurities in the electrolyte.

The advantages of the electrolytic methods include a short equilibration period and the facility with which the concentration of the calibration gas mixture can be changed by varying the current so that multiple-point calibration of analytical instruments can be readily carried out. The instruments can be small and portable. The disadvantages include the limited number of constituents which can be generated and the deviations from Faraday's law resulting from the evaporation of water or exhaustion of the electrolyte.

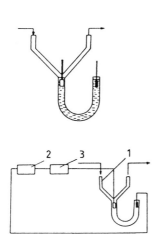

Fig. 16a. U-tube [4].
1 Electrolyser cell,
2 amperostat, 3 amperometer.

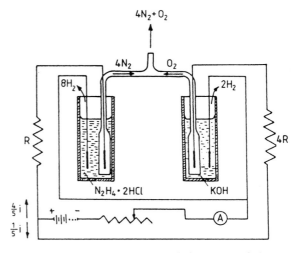

Fig. 16b. Apparatus to generate air by means of electrolysis ("electrolytic air") [4]

4.7 Chemical reactions

A gaseous constituent can be released from a precursor not only by electrolytic decomposition but also by thermal decomposition or by photochemical dissociation. For this purpose oxidative, reductive or catalytic chemical reactions can be used. One advantage of this method for the generation of calibration gas mixtures is the option of producing unstable, highly reactive gases like vinyl chloride, acrolein or acrylonitrile [2] by reaction from more stable precursors. Examples of the many possible reactions include: preparation of thiols by hydrolysis of S-alkylthiourea sulfates [66] or by thermal decomposition of dithiocarbamates [67], formation of HCl by sublimation from ammonium chloride and simultaneous collection of the NH_3 with an ion exchanger [68], formation of formaldehyde by deoligomerization of trioxane on a carborundum catalyst [69]. Ozone can be generated by the treatment of oxygen with UV light [70]. Nitrogen oxides are often formed in chemical reactions, for example, NO in the reaction of NH_3 on a platinum grid at 800 °C. In this case the rate of delivery of the ammonia can be controlled with a permeation tube [2]; NO_2 is generated by passing nitrogen oxide through an acidic permanganate solution. Formaldehyde is decomposed under UV light to yield CO and H_2 [71].

It must be taken into account that most reaction mechanisms are very complex. Sometimes two or more reactants must be present simultaneously in the gas phase before the required substance is formed. After an equilibration period, the reaction products should be formed stoichiometrically. It is often very difficult to remove products of secondary reactions or side reactions and unreacted precursors from the calibration gas mixture. It must also be remembered that the relationship between the reaction rate and the temperature is logarithmic.

5 Assessment of the reliability of a method

Characteristics such as precision, accuracy and reproducibility are important for the evaluation of a method or of the calibration gas mixture generated with the method. The accuracy of a method or the concentration of the calibration gas mixture is influenced by often undetected or unknown systematic errors. They may play a role during the generation of the calibration gas mixture or during the determination of the concentration by calculation or analysis. The requirements listed in Sections 2.1–2.4 must be met. The precision of a method is described by the mean variation which is determined as the standard deviation of the complete method. Besides the mean variation, the changes in a calibration gas mixture with time can be determined by repeated analysis. Repeated generation of the calibration gas mixture and its analysis yields the reproducibility of the method.

Interfering parameters (impurities in the complementary gas, substance losses resulting from leaks in the apparatus, etc.) and the constancy over a particular period must be known. Only the evaluation of the complete method of generating the calibration gas mixture, that is, the method with all operating procedures, can yield the required information as to the quality of the method, its susceptibility to interference, the cost in terms of staff and apparatus and the precision of the calibration gas mixtures it generates [5].

6 Summary

There are a variety of methods for the generation of calibration gas mixtures. This review has demonstrated that no standard method exists but that many methods with various advantages and disadvantages can be used; that is, there is no ideal method but many equivalent ones. The most suitable method for the particular application must be selected. The method of generation of a calibration gas mixture has to be decided separately for each situation. A method which has been applied successfully in one case may require changes for other applications or may be totally unsuitable so that it has to be replaced by another method. Before a method is chosen, the particular type of problem has to be analysed. Special requirements such as

- type of constituents,
- concentration,
- volume of the calibration gas mixture required,
- precision required, and
- stability

vary from application to application. The material required and the cost have to be taken into consideration too.

7 References

[1] *Verein Deutscher Ingenieure (VDI)* (1980) VDI-Richtlinie 3490 Blatt 1, Prüfgase – Begriffe und Erläuterungen. Beuth Verlag, Berlin.
[2] *Barratt RS* (1981) The Preparation of Standard Gas Mixtures. Analyst 106: 817–849.
[3] *Nelson GO* (1982) Controlled Test Atmospheres. Ann Arbor Science Publishers, Inc., Ann Arbor, Michigan.
[4] *Namiesnik J* (1984) Generation of the Standard Gaseous Mixtures. J Chromatogr 300: 79–108.
[5] *Becker WJ* (1977) Prüfgase und ihre Herstellung (Teil 1–3). Staub-Reinhalt Luft 37: 183–188, 278–283, 426–431.
[6] *Roessel H, Buchholz N, Hartkamp H* (1983) Membrandosierer zur Herstellung sekundärer Standards als Eich- und Prüfgase in der Gasspurenanalyse. Fresenius Z Anal Chem 316: 142–149.
[7] *Verein Deutscher Ingenieure (VDI)* (1980) VDI-Richtlinie 3490 Blatt 2, Prüfgase – Herstellungsverfahren-Übersicht. Beuth Verlag, Berlin.
[8] *Verein Deutscher Ingenieure (VDI)* (1980) VDI-Richtlinie 3490, Blatt 3, Prüfgase – Anforderungen und Maßnahmen für den Transfer. Beuth Verlag, Berlin.

[9] *Verein Deutscher Ingenieure (VDI)* (1980) VDI-Richtlinie 3490 Blatt 4, Prüfgase – Herstellung mit gravimetrischen Methoden. Beuth Verlag, Berlin.

[10] *Verein Deutscher Ingenieure (VDI)* (1980) VDI-Richtlinie 3490 Blatt 5, Prüfgase – Ermittlung der Zusammensetzung durch Vergleichsmethoden. Beuth Verlag, Berlin.

[11] *Verein Deutscher Ingenieure (VDI)* (1980) VDI-Richtlinie 3490 Blatt 6, Prüfgase – Dynamische Herstellung mit Gasmischpumpen. Beuth Verlag, Berlin.

[12] *Verein Deutscher Ingenieure (VDI)* (1980) VDI-Richtlinie 3490 Blatt 7, Prüfgase – Dynamische Herstellung durch periodische Injektion. Beuth Verlag, Berlin.

[13] *Verein Deutscher Ingenieure (VDI)* (1981) VDI-Richtlinie 3490 Blatt 8, Prüfgase – Herstellung durch kontinuierliche Injektion. Beuth Verlag, Berlin.

[14] *Verein Deutscher Ingenieure (VDI)* (1980) VDI-Richtlinie 3490 Blatt 9, Prüfgase – Herstellung durch Permeation der Beimengungen in einen Grundgasstrom. Beuth Verlag, Berlin.

[15] *Verein Deutscher Ingenieure (VDI)* (1981) VDI-Richtlinie 3490 Blatt 10, Prüfgase – Herstellen von Prüfgasen durch Mischen von Volumenströmen-Kapillardosierer. Beuth Verlag, Berlin.

[16] *Verein Deutscher Ingenieure (VDI)* (1980) VDI-Richtlinie 3490 Blatt 11, Prüfgase – Herstellung nach der volumetrisch-statischen Methode unter Verwendung von Kunststoffbeuteln. Beuth Verlag, Berlin.

[17] *Verein Deutscher Ingenieure (VDI)* (1988) VDI-Richtlinie 3490 Blatt 12, Prüfgase – Herstellung von Prüfgasen durch manometrische Methoden. Beuth Verlag, Berlin.

[18] *Verein Deutscher Ingenieure (VDI)* (1992) VDI-Richtlinie 3490 Blatt 13, Prüfgase – Herstellen von Prüfgasen durch Sättigungsmethoden. Beuth Verlag, Berlin.

[19] *Verein Deutscher Ingenieure (VDI)* (1994) VDI-Richtlinie 3490 Blatt 14, Prüfgase – Herstellen von Prüfgasen nach der volumetrisch-statischen Methode unter Verwendung von Glasbehältern. Beuth Verlag, Berlin.

[20] *Verein Deutscher Ingenieure (VDI)* (1985) VDI-Richtlinie 3490 Blatt 15, Prüfgase – Direkte Ermittlung der Beimengung eines Prüfgases durch Gasdichtemessung-Gasdichtewaage. Beuth Verlag, Berlin.

[21] *Verein Deutscher Ingenieure (VDI)* (1994) VDI-Richtlinie 3490 Blatt 16, Prüfgase – Herstellen von Prüfgasen mit Blenden-Mischstrecken. Beuth Verlag, Berlin.

[22] *International Standard Organization (ISO)* (1981) ISO 6144, Gas analysis, Preparation of calibration gas mixtures, Static volumetric methods. Geneva, Beuth Verlag, Berlin.

[23] *International Standard Organization (ISO)* (1986) ISO 6145-3, Gas analysis, Preparation of calibration gas mixtures, Dynamic volumetric methods, Part 3, Periodic injections into a flowing gas stream. Geneva, Beuth Verlag, Berlin.

[24] *International Standard Organization (ISO)* (1986) ISO 6145-4, Gas analysis, Preparation of calibration gas mixtures, Dynamic volumetric methods, Part 4, Continuous injection method, Geneva, Beuth Verlag, Berlin.

[25] *International Standard Organization (ISO)* (1986) ISO 6145-5, Gas analysis, Preparation of calibration gas mixtures, Dynamic volumetric methods, Part 5, Capillary calibration devices. Geneva, Beuth Verlag, Berlin.

[26] *International Standard Organization (ISO)* (1986) ISO 6145-6, Gas analysis, Preparation of calibration gas mixtures, Dynamic volumetric methods, Part 6, Sonic orifices. Geneva, Beuth Verlag, Berlin.

[27] *International Standard Organization (ISO)* (1979) ISO 6146, Gas analysis, Preparation of calibration gas mixtures; Manometric method. Geneva, Beuth Verlag, Berlin.

[28] *International Standard Organization (ISO)* (1979) ISO 6147 Gas analysis, Preparation of calibration gas mixtures, Saturation methods. Geneva, Beuth Verlag, Berlin.

[29] *Eaton WC* (1978) Use of the flame photometric detector method for measurement of sulfur dioxide in ambient air a technical assistentance document. Report No EPA-600/4-78-024.

[30] *Roth M, Vejrosta J, Novák J* (1983) Effect of gas-phase non-ideality on the concentration of standard gaseous mixtures prepared by the saturation method. J Chromatogr 262: 305–310.

[31] *Drägerwerk* (1991) Dräger Probenahme-Handbuch, Lübeck.

[32] *Grosjean D* (1985) Wall loss of gaseous pollutants in outdoor teflon chambers. Environ Sci Technol 19: 1059–1065.

[33] *Denyszyn RB, Sassaman T* (1987) Parts-per-billion gaseous mixtures: a new challenge. In: Taylor JK (Ed) Sampling and calibration for atmospheric measurements. ASTM STP 957. American Society for Testing and Materials, Philadelphia, 101–109.

[34] *Greenhouse S, Andrawes F* (1990) Generation of gaseous standards using exponential dilution flasks in series. Anal Chim Acta 236: 221–226.

[35] *Wilde K* (1992) Verfahren zur Herstellung von Prüfgasen. GIT Fachz Lab 2: 930–931.

[36] *Schmicker D* (1988) Herstellung von Prüfgasen für die Spurenanalyse und ihre Anwendung. Gwf-gas/erdgas 129: 38–41.

[37] *Howe GB, Albritton JR, Tompkins SB, Jayanty RKM, Decker CE, von Lehmden DJ* (1989) Stability of parts per-million organic cylinder gases and results of source test analysis audits. Status Report Number 10. Report No EPA-600/3-89-017.

[38] *Hughes EE, Davenport AJ, Woods PT, Zielinski WL* (1991) Intercomparison of a range of primary gas standards of carbon monoxide in nitrogen and carbon dioxide in nitrogen prepared by the National Institute of Standards and Technology and the National Physical Laboratory. Environ Sci Technol 25: 671–676.

[39] *Hughes EE, Gilkey RK, Scawin JH, Frankvoort W, Chain D, Hamon B, Uchikawa H, Yano H, Fukuchi T, Kurihara C* (1987) International comparison of reference gases for calibrating measuring instruments of air pollutants. J Japan Soc Air Pollut 22: 460–472.

[40] *Rhoderick GC* (1991) Development of a gas standard reference material containing eighteen volatile organic compounds. Fresenius J Anal Chem 341: 524–531.

[41] *Baumer D, Hamm R, Riedel E* (1984) Die wichtigsten Herstellungsmethoden für Gasgemische. Gwf-gas/erdgas 125: 208–213.

[42] *Moore GS, Farmer R* (1987) Increasing calibration accuracy of a gas blending device. Air Pollution Control Association, Proceedings, 80th Annual Meeting 4, 87-65.8: 2–14.

[43] *Goldan PD, Kuster WC, Albritton DL* (1986) A dynamic dilution system for the production of sub-ppb concentrations of reactive and labile species. Atmos Environ 20: 1203–1209.

[44] *Heidrich K* (1981) Präzise Herstellung von Prüfgasen für die Analysen-Technik. Chemie-Technik 9: 889–891.

[45] *Tschickardt M* (1989) Routineeinsatz des Thermodesorbers ATD-50 in der Gefahrstoffanalytik. Perkin Elmer Heft 48, Perkin Elmer (Ed), Überlingen.

[46] *Leichnitz K* (1983) Generating calibration gas mixtures for calibration of methods used in environmental analysis. Pure Appl Chem 55: 1239–1250.

[47] *Hartkamp H* (1973) Kapillardosierer für die Herstellung primärer Standards zu Eich- und Prüfzwecken. Schriftenreihe der Landesanstalt für Immisions-und Bodennutzungsschutz des Landes Nordrhein-Westfalen 29: 69–70.

[48] *Goelen E, Rymen T* (1992) The generation of VOCs in air: A New Approach to the Capillary Dosage Technique. In: Brown RH, Curtis M, Saunders KJ, Vandendriessche S (Ed): Clean air at work-new trends in assessment and measurement for the 1990s. Proceedings of an International Symposium held in Luxembourg, Sept 9–13, 1991, Royal Society of Chemistry, London.

[49] *Namiesnik J* (1983) Permeation devices for the preparation of standard gaseous mixtures. Chromatographia 17: 47–48.

[50] *Peperstraete HJ* (1982) Gaseous reference materials-certification-traceability. In: Versino B, Ott H (Ed) Physico-chemical behaviour of atmospheric pollutants. Proceedings of the Second European Symposium held in Varese, Italy, Sept 29–Oct 1, 1981, Reidel Publishing Company, London, 68–75.

[51] *Farwell SO, Kagel RA, Barinaga CJ, Goldan PD, Kuster WC, Fehsenfeld FC, Albritton D* (1987) Intercomparison of two techniques for the preparation of gaseous sulfur calibration standards in the low sub-ppb range. Atmos Environ 21: 1983–1987.

[52] *Müller RE, Schurath U* (1983) Generation of formaldehyde in test atmospheres with low concentrations of hydrogen and carbon monoxide. Anal Chem 55: 1440–1442.

[53] *Schoene K, Steinhanses J* (1989) Generating vapour mixtures for calibration purposes, II. Dynamic diffusive system. Fresenius Z Anal Chem 335: 557–561.

[54] *Hollander JCT, Nielen MWF* (1986) Feasibility studies tenax adsorption tubes. Report R 87/22, TNO, Nutrition and Food Research Institute, Zeist.

[55] *Pitombo LRM, Cardoso AA* (1990) Standard gas mixture production based on the diffusion method. Intern J Environ Anal Chem 39: 349–360.

[56] *Hufschmidt R* (1982) Erzeugung von Prüfgasen im Labor mittels Kapillardiffusion. Staub-Reinhalt Luft 42: 390–391.

[57] *Worthington B, Rey AR* (1991) Generation of dynamic standard test atmospheres for aromatic compounds by using the diffusion vial method. Am Ind Hyg Assoc J 52: 464–468.

[58] *Glaser RA, Woodfin WJ* (1981) A method for sampling and analysis of 2-nitropropane in air. Am Ind Hyg Assoc J 42: 18–22.

[59] *Kettrup A* (1982) Apparatur zur Herstellung von Prüfgasen. GIT Fachz Lab 26: 556–558.

[60] *Anderson CC, Gunderson EC, Coulson DM* (1981) Sampling and analytical methodology for workplace hazards. State of the Art and Future Trends. ACS Symposium Series No 149: 3–19.

[61] *Lide DR* (Ed) (1997) CRC Handbook of chemistry and physics. 78th Edition, CRC Press. Inc, Boca Raton, Florida.

[62] *D'Ans J, Lax E* (1983) Taschenbuch für Chemiker und Physiker. Band 2, 4. Aufl, Springer-Verlag, Heidelberg-Berlin.

[63] *Landolt H, Börnstein R* Zahlenwerte und Funktionen aus der Physik, Chemie, Astronomie, Geophysik und Technik. Springer-Verlag, Heidelberg-Berlin.

[64] *Green DW* (Ed) (1997) Perrys' chemical engineers' handbook, 7th Edition, McGraw-Hill Book Company, New York.

[65] *Flammenkamp E, Ludwig E, Kettrup A* (1992) Reference material: Active charcoal tubes charged with dynamically generated multi component test gas of the three analytes benzene, toluene and m-xylene. In: Brown RH, Curtis M, Saunders KJ, Vandendriessche S (Ed) Clean air at work-new trends in assessment and measurement for the 1990s. Proceedings of an International Symposium held in Luxembourg, Sept 9–13, 1991, Royal Society of Chemistry, London.

[66] *Ishikawa K, Hobo T, Suzuki S* (1984) Generation of trace amounts of alkanethiol standard gases using reaction gas chromatography. J Chromatogr 295: 445–452.

[67] *Konieczka P, Namiesnik J, Biernat JF* (1991) Generation of standard gaseous mixtures by thermal decomposition of surface compounds. J Chromatogr 540: 449–455.

[68] *Lindgren PF* (1991) Trace level gaseous hydrochloric acid measurement standard based on sublimation of ammonium chloride. Anal Chem 63: 1008–1011.

[69] *Geisling KL, Miksch RR, Rappaport SM* (1982) Generation of dry formaldehyde at trace levels by the vapour-phase depolymerization of trioxane. Anal Chem 54: 140–142.

[70] *Dorko WD, Hughes EE* (1987) Special calibration systems for reactive gases and other difficult measurements. In: Taylor JK (Ed) Sampling and calibration for atmospheric measurements. ASTM STP 957. American Society for Testing and Materials, Philadelphia, 132–137.

[71] *Heidrich K* (1990) Prüfgasgeneratoren für Kohlenmonoxid und Stickstoffdioxid. WLB Wasser, Luft Boden 10: 54.

Authors: *E. Flammenkamp, U. Risse*

Analytical Methods

Acrylates (Methyl acrylate, Ethyl acrylate, Butyl acrylate)

Method number   1

Application   Air analysis

Analytical principle Gas chromatography

Completed in   July 1992

Summary

For the determination of airborne concentrations of acrylates known air volumes are drawn through an activated carbon tube by means of a sampling pump. The adsorbed acrylates are eluted with carbon disulfide. After gas chromatographic separation from hydrocarbons and other airborne constituents they are analysed by flame ionization gas chromatography. The result is calculated using an internal standard (n-tetradecane) which is added to the desorption solution.

Precision:
Standard deviation (rel.) $s = 1.5\%$
Mean variation $u = 3.5\%$
at a concentration of methyl acrylate and ethyl acrylate of 5 mL/m^3 and butyl acrylate of 10 mL/m^3 in air and $n = 10$ determinations

Detection limit:
0.1 mL/m^3 for methyl acrylate and ethyl acrylate (equivalent with 0.04 mg/m^3) and 0.05 mL/m^3 (equivalent with 0.2 mg/m^3) butyl acrylate referring to a sample volume of 25 L

Recovery rate:
$\eta = 0.84$ (84%) for methyl acrylate
$\eta = 0.87$ (87%) for ethyl acrylate
$\eta = 0.90$ (90%) for butyl acrylate

Recommended sampling time: 8 h
Recommended sample volume: 25 L

Methyl acrylate

$CH_2=CHCOOCH_3$

is a colourless liquid (molecular weight 86.09 g/mole, boiling point 80 °C, vapour pressure 93 hPa at 20 °C, density 0.956 g/cm^3) with a perceptible odour. It is strongly irritant to the skin, the eyes and the respiratory tract. The currently valid MAK value (1998) is 5 mL/m^3 and 18 mg/m^3 [1].

Ethyl acrylate

$CH_2=CHCOOC_2H_5$

is a colourless liquid (molecular weight 100.12 g/mole, boiling point 99 °C, vapour pressure 39 hPa at 20 °C, density 0.924 g/cm^3) with a perceptible odour. It is strongly irritant to the skin, the eyes and the respiratory tract. The currently valid MAK value (1998) is 5 mL/m^3 and 21 mg/m^3 [1].

Butyl acrylate

$CH_2=CHCOO(CH_2)_3CH_3$

is a colourless liquid (molecular weight 128.17 g/mole, boiling point 145 °C, vapour pressure 5.3 hPa at 20 °C, density 0.894 g/cm^3) with a perceptible odour. It is strongly irritant to the skin, the eyes and the respiratory tract. The currently valid MAK value (1998) is 2 mL/m^3 and 11 mg/m^3 [1].

The acrylates are used in the production of polymers, polymer dispersions and lacquer raw materials. They polymerise easily with light, heat or peroxides as catalysts.

Authors: *W. Merz, W. Krämer*
Examiner: *H. Muffler*

Acrylates (Methyl acrylate, Ethyl acrylate, Butyl acrylate)

Method number	1
Application	Air analysis
Analytical principle	Gas chromatography
Completed in	July 1992

Contents

1 General principles

For the determination of airborne concentrations of acrylates known air volumes are drawn through an activated carbon tube by means of a sampling pump. The adsorbed acrylates are eluted with carbon disulfide. After gas chromatographic separation from hydrocarbons and other airborne constituents they are analysed by flame ionization gas chromatography. The result is calculated using an internal standard (*n*-tetradecane) which is added to the desorption solution.

2 Equipment, chemicals and solutions

2.1 Equipment

Pump equipped with flow control or gasmeter and holder for the activated carbon tubes
Soap bubble flowmeter
Stopwatch
Adsorption tubes filled with activated carbon, standardized, containing two activated carbon fillings of about 100 mg and 50 mg separated from each other by an inert porous polymeric material
10 mL and 100 mL Volumetric flasks
Septum vials
Glass cutter
1 mL Dosing pipette
Microliter syringes or microcaps
Gas chromatograph equipped with flame ionization detector, suitable for temperature-programmed operation and with split/splitless injector
Recorder or integrator

2.2 Chemicals

Carbon disulfide, analytical grade, e. g. from Merck
Methyl acrylate, 99% purity, e. g. from Aldrich
Ethyl acrylate, 99% purity, e. g. from Aldrich
Butyl acrylate, 99% purity, e. g. from Aldrich
n-Tetradecane 99.8% purity, e. g. from Riedel de Haen

2.3 Solutions

Methyl acrylate stock solution:
500 µL ($\cong 478$ mg) methyl acrylate are transferred into a 100 mL volumetric flask and diluted to the mark with carbon disulfide with occasionally shaking to prepare the stock solution. The solution contains 4.78 g/L methyl acrylate.

Ethyl acrylate stock solution:
500 µL ($\cong 462$ mg) ethyl acrylate are transferred into a 100 mL volumetric flask and diluted to the mark with carbon disulfide with occasionally shaking to prepare the stock solution. The solution contains 4.42 g/L ethyl acrylate.

Butyl acrylate stock solution:
1500 μL ($\cong$1341 mg) butyl acrylate are transferred into a 100 mL volumetric flask
and diluted to the mark with carbon disulfide with occasionally shaking. The solution
contains 13.41 g/L butyl acrylate.

The stock solutions can be stored in a refrigerator longer than half a year.

n-Tetradecane stock solution (internal standard):
10 μL ($\cong$7.61 mg) *n*-tetradecane are pipetted into a 100 mL volumetric flask containing 50 mL carbon disulfide to prepare the *n*-tetradecane stock solution. Then the solution is diluted to the mark with carbon disulfide. The solution contains 76.1 mg/L *n*-tet-
radecane.

Desorption solution:
10 mL of the *n*-tetradecane stock solution are transferred into a 100 mL volumetric
flask and diluted up to the mark with carbon disulfide. The *n*-tetradecane content of
the solution is 7.61 μg/mL *n*-tetradecane.

Methyl acrylate calibration solutions:
Volumes of 1 mL each of the *n*-tetradecane stock solution and 5, 10, 50, 100, 1000 and
2000 μL of the methyl acrylate stock solution are transferred into six different 10 mL
volumetric flasks and diluted to the mark with carbon disulfide (cf. Tab. 1).
The *n*-tetradecane content of each solution is 7.61 μg/mL. The methyl acrylate content
is 2.39, 4.78, 23.9, 47.8, 478 and 956 μg/mL. By use of these solutions methyl acrylate
concentrations can be determined in the range of 96–38 160 μg/m^3 for a sample air volume of 25 L.

Table 1. Pipetting scheme for the preparation of methyl acrylate calibration standards.

Volume of the methyl acrylate stock solution μL	Volume of the *n*-tetradecane stock solution mL	Total volume of the calibration standard mL	Concentration of *n*-tetradecane μg/mL	Concentration of methyl acrylate	
				μg/mL	μg/m^3
5	1	10	7.61	2.39	95.6
10	1	10	7.61	4.78	191.2
50	1	10	7.61	23.9	956
100	1	10	7.61	47.8	1912
1000	1	10	7.61	478	19120
2000	1	10	7.61	956	38240

Ethyl acrylate calibration solutions:
Volumes of 1 mL each of the *n*-tetradecane stock solution and 5, 10, 50, 100, 1000 and
2000 μL of the ethyl acrylate stock solution are transferred into six different 10 mL volumetric flasks and diluted to the mark with carbon disulfide (cf. Tab. 2).
The *n*-tetradecane content of each solution is 7.61 μg/mL. The ethyl acrylate content is
2.21, 4.42, 22.1, 44.2, 442 and 884 μg/mL. By use of these solutions ethyl acrylate concentrations can be determined in the range of 88.4–35 360 μg/m^3 for a sample air volume of 25 L.

Table 2. Pipetting scheme for the preparation of ethyl acrylate calibration standards.

Volume of the ethyl acrylate stock solution µL	Volume of the n-tetradecane stock solution mL	Total volume of the calibration standard mL	Concentration of n-tetradecane µg/mL	Concentration of ethyl acrylate	
				µg/mL	µg/m³
5	1	10	7.61	2.21	88.4
10	1	10	7.61	4.42	176.8
50	1	10	7.61	22.1	884
100	1	10	7.61	44.2	1768
1000	1	10	7.61	442	17680
2000	1	10	7.61	884	35360

Butyl acrylate calibration solutions:
Volumes of 1 mL each of the n-tetradecane stock solution and 1, 5, 10, 50, 100, 1000 and 2000 µL of the butyl acrylate stock solution are transferred into seven different 10 mL volumetric flasks and diluted up to the mark with carbon disulfide (cf. Tab. 3). The n-tetradecane content of each solution is 7.61 µg/mL. The butyl acrylate content is 1.34, 6.71, 13.41, 67.1, 134.1, 1341 and 2682 µg/mL. By use of these solutions butyl acrylate concentrations can be determined in the range of 53.64–107280 µg/m³ for a sample air volume of 25 L.

Table 3. Pipetting scheme for the preparation of butyl acrylate calibration standards.

Volume of the butyl acrylate stock solution µL	Volume of the n-tetradecane stock solution mL	Total volume of the calibration standard mL	Concentration of n-tetradecane µg/mL	Concentration of butyl acrylate	
				µg/mL	µg/m³
1	1	10	7.61	1.34	53.64
5	1	10	7.61	6.71	268.2
10	1	10	7.61	13.41	536.4
50	1	10	7.61	67.1	2682
100	1	10	7.61	134.1	5364
1000	1	10	7.61	1341	53640
2000	1	10	7.61	2682	107280

3 Sample collection and preparation

At the sampling location an activated carbon tube is opened with a glass cutter and connected to the personal air sampling pump. Pump and tube are worn by a person during the working hours or they can be used in a static position. The flow rate is about 3 L/h. For an eight hour sampling period this corresponds to a sample volume of about 25 L.

The counting device of the pump has to be calibrated before and after the sampling under the same flow rate conditions. This is carried out by use of a soap bubble counter and a stopwatch.

After sampling the contents of the loaded sampling tube are transferred to a septum vial. 1 mL of the desorption solution is added. After the vial has been treated in an ultrasonic bath for 5 min the liquid phase is separated and transferred into another vial because the acrylates decompose if they are in contact with activated carbon and carbon disulfide.

A blank value is determined for each analysis series.

4 Operating conditions for gas chromatography

Apparatus:	Gas chromatograph equipped with flame ionization detector, autosampler and dual capillary system	
Columns:	Material:	Quartz capillary, fused silica
	Length:	30 m
	Internal diameter:	0.259 mm,
Stationary phase:	DB-1 (100% dimethylpolysiloxane, crosslinked and chemically bound)	
	Film thickness:	0.5 μm
	DB-WAX (polyethylene glycol, crosslinked and chemically bound)	
	Film thickness:	0.5 μm
Detector:	Flame ionization detector	
Temperatures:	Injector:	210 °C
	Detector:	255 °C
	Oven:	multistage temperature program
		isothermal 35 °C, 3 min
		3 °C/min up to 65 °C
		isothermal 65 °C, 0 min
		6 °C/min up to 120 °C
		isothermal 120 °C, 3 min
		10 °C/min up to 220 °C
		isothermal 220 °C, 5 min
Carrier gas:	Helium:	2 mL/min
Detector gases:	Synthetic air	
	Hydrogen	
Injection volume:	1 μL	

5 Analytical determination

1 µL of the desorption solution is injected into the gas chromatograph under the described conditions. After the gas chromatographic separation the acrylates are detected by a flame ionization detector (cf. Fig. 1 and Fig. 2).

6 Calibration

1 µL of each of the calibration solutions (cf. Sect. 2.3) is injected into the gas chromatograph. The calibration curve is obtained by plotting the peak areas (peak heights) against the quantities of the acrylates of the corresponding calibration solution. The linearity of the curve is checked.

The calibration factor f_A (response factor) is determined by means of the peak areas or peak heights obtained for n-tetradecane and the acrylates of the different dilutions according to:

$$f_A = \frac{F_T \cdot m_A}{F_A \cdot m_T}$$

Legend:

f_A Calibration factor for acrylates
F_T Peak area (or peak height) of the n-tetradecane in the acrylate/n-tetradecane solution
F_A Peak area (or peak height) of the acrylate in the acrylate/n-tetradecane solution
m_A Weight of the acrylate in 1 mL of the acrylate/n-tetradecane solution in µg
m_T Weight of n-tetradecane in 1 mL acrylate/n-tetradecane solution in µg.

The value of f_A is nearly the same for all dilutions. The average value $\bar{f}_A$ is used for the calculation of the analytical result.

7 Calculation of the analytical result

The concentration of the airborne acrylates in in mg/m^3 is calculated as follows:

$$\rho = \frac{F_A \cdot m_T \cdot \bar{f}_A \cdot 1000}{F_T \cdot V}$$

Legend:

ρ Concentration by weight of the airborne acrylates in mg/m^3
F_A Peak area (or peak height) of the airborne acrylates for the sample

F_T Peak area (or peak height) of the *n*-tetradecane in the sample
m_T *n*-Tetradecane in 1 mL *n*-tetradecane calibration solution in mg
V Sample volume in L
$\bar{f}_A$ Average calibration factor for acrylates

The calculation of the concentration by volume σ in mL/m^3 of the methyl acrylate from ρ at 20 °C and 1013 hPa is calculated as follows:

$$\sigma = 0.28 \cdot \rho$$

The calculation of the concentration by volume σ in mL/m^3 of the ethyl acrylate from ρ at 20 °C and 1013 hPa is calculated as follows:

$$\sigma = 0.24 \cdot \rho$$

The calculation of the concentration by volume σ in mL/m^3 of the butyl acrylate from ρ at 20 °C and 1013 hPa is calculated as follows:

$$\sigma = 0.19 \cdot \rho$$

Legend:

σ Concentration by volume of the acrylates in the sample air in mL/m^3 (ppm)

8 Reliability of the method

The method was validated with the following MAK values (1991): methyl acrylate 5 mL/m^3, ethyl acrylate 5 mL/m^3, butyl acrylate 10 mL/m^3.

8.1 Precision

The precision of the method was determined ten times at concentrations of a tenth of the MAK, at the MAK and at the twofold MAK by transferring corresponding solutions into a glass tube. At a flow rate of 2.8 L/h a volume of 22.4 L purified air was drawn through the glass tube and two activated carbon tubes connected in series. For the determination of the three esters a relative standard deviation of 1.5% and a mean variation of 3.5% was obtained.

8.2 Recovery rate

The efficiency of the collection of the three esters on the activated carbon tubes was tested with the same arrangement and the same dosage quantities. The average recovery rates were: for methyl acrylate 84%, for ethyl acrylate 87% and for the butyl acrylate 90%.

Under the described conditions the humidity did not influence the measuring result.

8.3 Detection limit

The absolute detection limit is 10 ng for methyl acrylate and ethyl acrylate and 5 ng for the butyl acrylate. The relative detection limit is 400 $\mu g/m^3$ for methyl acrylate and ethyl acrylate and 200 $\mu g/m^3$ for the butyl acrylate at a sample volume of 25 L, 1 mL sample solution and 1 μL injection volume.

8.4 Sources of error

Interferences were not observed.

9 Discussion

This method can be used for personal air sampling as well as stationary sampling in the assessment of work areas.

For a period of 5 minutes (time criterion of the short-term exposure limit) at a flow rate of 600 mL/min sample air is drawn through the sampling tube by use of a sampling pump to determine the adherence of the peak concentrations (the mentioned acrylates are listed in the peak limitation category I).

The shelf life of the adsorbed acrylic acid esters is at least 7 days.

The selectivity has to be tested in each individual case.

Apparatus: Hewlett Packard 5880 equipped with FID and autosampler
             Hewlett Packard 7672 A, dual capillary system

10 References

[1] *Deutsche Forschungsgemeinschaft* (1998) List of MAK and BAT values 1998. Maximum concentrations and biological tolerance values at the workplace. Report No 34 by the Commission for the Investigation of Health Hazards of Chemical Compounds in the work area. WILEY-VCH Verlag GmbH, Weinheim

Authors: *W. Merz, W. Krämer*
Examiner: *H. Muffler*

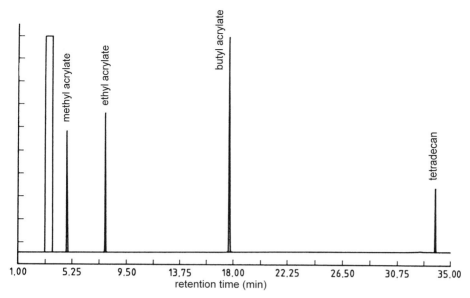

Fig. 1. Chromatogram of the acrylates on an unpolar column (DB1).

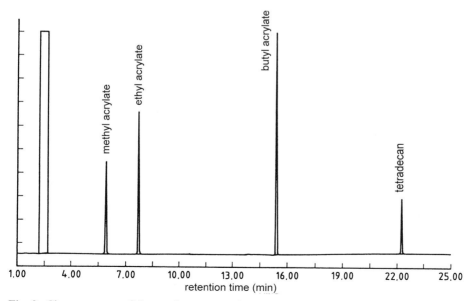

Fig. 2. Chromatogram of the acrylates on a polar column (DBWAX).

2-Chloroethanol

Method number	1
Application	Air analysis
Analytical principle	Gas chromatography
Completed in	April 1993

Summary

A measured air volume is drawn through a silica gel tube by use of a sampling pump. The adsorbed 2-chloroethanol is desorbed with acetone and determined by flame ionization gas chromatography.
Quantitative analysis is carried out using an internal standard method and peak areas.

Precision:	Standard deviation (rel.) $s = 3.7\%$
	Mean variation $u = 8\%$
	at a concentration of 0.48 mg/m^3
	and $n = 10$ determinations
Quantification limit:	0.06 mg 2-chloroethanol per m^3 air
	(referring to a sample volume of 25 L)
Recovery rate:	$\eta = 0.95$ (95%)
Recommended sampling time:	8 h
Recommended sample volume:	25 L

2-Chloroethanol

$ClCH_2-CH_2OH$

2-Chloroethanol (ethylene chlorohydrin) is a colourless liquid with a molecular weight of 80.52 g/mole, a density of 1.20 g/mL and a boiling point of 129 °C. It is miscible with water and alcohols.
2-Chloroethanol is used as a solvent for acetylcellulose, acids and basic pigments. It is also an intermediate product in the synthesis of ethylene oxide, insecticides and plasticizers.

2-Chloroethanol is irritant to the skin and mucuous membranes. The risk of skin re-sorption also exists. Phosgene can be formed when heating the substance.
The currently valid MAK value (1998) is 3,3 mg/m^3 and 1 mL/m^3 [1].

Authors: *J. Oldeweme, W. Merz*
Examiners: *A. Kettrup, H. Weber*

2-Chloroethanol

Method number   1

Application   Air analysis

Analytical principle Gas chromatography

Completed in   April 1993

Contents

1 General principles

A measured air volume is drawn through a silica gel tube by use of a sampling pump. The adsorbed 2-chloroethanol is desorbed with acetone and determined by flame ionization gas chromatography.

Quantitative analysis is carried out using an internal standard method and peak areas.

2 Equipment, chemicals and solutions

2.1 Equipment

Gas chromatograph equipped with flame ionization detector
Recorder or integrator
Adsorption tubes (length 7 cm, outer diameter 0.6 cm), consisting of two silica gel sections of about 100 mg main section and 50 mg back-up section which are separated from each other by porous polymeric material
Suction pump, capable of operating in the range about 3–5 L/h
Gasmeter
Thermometer
Barometer
10 mL and 100 mL Volumetric flasks
Sample vials with polytetrafluoroethylene (PTFE) coated septum and screw closure cap
Glass cutter
Glass capillaries

2.2 Chemicals

2-Chloroethanol, > 99% purity
Acetone, analytical grade
n-Dodecane, analytical grade

2.3 Solutions

2-Chloroethanol stock solution:
500 µL ($\cong$ 600 mg) of 2-chloroethanol is transferred into a 100 mL volumetric flask and diluted up to the mark with acetone with occasionally shaking.

n-Dodecane stock solution:
50 mL acetone is transferred into a 100 mL volumetric flask, 10 µL (7.50 mg) n-dodecane added, and diluted up to the mark with acetone.

Desorption solution:
10 mL of the n-dodecane solution is transferred into a 100 mL volumetric flask and diluted up to the mark with acetone. This solution contains 7.5 µg/mL n-dodecane.

2-Chloroethanol/n-dodecane calibration solution:
A volume of each 1 mL of the n-dodecane stock solution and 5, 10, 20, 50 and 100 µL of the 2-chloroethanol stock solution are transferred into five separate 10 mL volumetric flasks and diluted up to the mark with acetone (cf. Tab. 1). Using these solutions, a concentration range of 0.12–2.40 mg/m^3 of 2-chloroethanol for a sample air volume of 25 L can be detected.

Table 1. Pipetting scheme for the preparation of calibration standards.

Volume of the stock solution	Total volume of the calibration standard	Concentration of 2-chloroethanol	Concentration of 2-chloroethanol for a sample air volume of 25 L
μL	mL	mg/L	mg/m³
5	10	3	0.12
10	10	6	0.24
20	10	12	0.48
50	10	30	1.20
100	10	60	2.40

3 Sample collection and preparation

Using a flow stabilized pump or a pump which is controlled by a gasmeter, the sample air is drawn through the adsorbent tube at a flow rate in the range 3–5 L/h. The total sample air volume should not exceed 25 L.
The decisive parameters for the concentration determination like sample volume, temperature and ambient pressure at the measuring location have to be determined.
The loaded tubes are closed with plastic caps. They can be stored up to one week.
For the desorption of the 2-chloroethanol the contents of a loaded tube are transferred into a sample vial, 1 mL of desorption solution (cf. Sect. 2.3) added and the vial is shaken occasionally. After a desorption time of 1 hour, the sample can be analysed by gas chromatography.

4 Operating conditions for gas chromatography

Column: Material: Quartz capillary, fused silica
                  Length: 30 m
                  Internal diameter: 0.25 mm
Stationary phase: DB 1701 (14%)-Cyanopropylphenyl-(86%)-dimethylsiloxane
                  copolymer, crosslinked and chemically bound)
Detector: Flame ionization detector
Temperatures: Injector: 210 °C
                  Detector: 250 °C
                  Oven: Start at 40 °C for 10 min
                                       with 6 °C/min up to 180 °C
                                       10 min at 180 °C
Carrier gas: Helium: 2 mL/min

Detector gases:	Hydrogen:	30 mL/min
	Synthetic air:	360 mL/min
Injection volume:	1 µL	

5 Analytical determination

The operating parameters for the gas chromatography are adjusted. Replicate volumes of 1 µl are injected by means of a microliter syringe.

6 Calibration

Volumes of 1 µl of each calibration solution (cf. Sect. 2.3) are injected into the gas chromatograph. The peak areas of the 2-chloroethanol and n-dodecane internal standard are measured using a recorder and/or integrator. Determine the peak responses of the 2-chloroethanol relative to those of the internal standard. To obtain the calibration curve these relative responses are plotted against the 2-chloroethanol concentrations in the calibration solutions.

7 Calculation of the analytical result

The weights of 2-chloroethanol (in mg) are determined from the relative peak responses of 2-chloroethanol and n-dodecane and the calibration curve prepared in Sect. 6. The concentration by weight ρ (mg 2-chloroethanol/m^3 air) is calculated as follows:

$$\rho = \frac{X}{V_Z \cdot \eta} \cdot \frac{273 + t_g}{273 + t_a}$$

At 20 °C and 1013 hPa:

$$\rho_0 = \rho \, \frac{273 + t_a}{293} \cdot \frac{1013\,\text{hPa}}{p_a}$$

The corresponding concentration by volume (independent of the pressure and temperature) is:

$$\sigma = \rho_0 \, \frac{24.1\,\text{L/mole}}{80.5\,\text{g/mole}}$$

$$\sigma = \rho \cdot \frac{273 + t_a}{p_a} \cdot \frac{1013\,\text{hPa}}{293} \cdot \frac{24.1\,\text{L/mole}}{80.5\,\text{g/mole}}$$

$$\sigma = \rho \cdot \frac{273 + t_a}{p_a}\,1.03\,\frac{\text{hPa} \cdot \text{mL}}{\text{mg}}$$

At $t_a = 20\,°\text{C}$ and $p_a = 1013\,\text{hPa}$:

$$\sigma = \rho \cdot 0.33\,\frac{\text{mL}}{\text{mg}}$$

Legend:

X Weight of 2-chloroethanol in mg in the sample solution
V_Z Sample volume in m^3
η Recovery rate
t_g Temperature in the gasmeter in $°\text{C}$
t_a Temperature of the ambient air in $°\text{C}$
p_a Ambient pressure in hPa
ρ Concentration by weight of airborne 2-chloroethanol in mg/m^3 referring to t_a and p_a (see above)
ρ_0 Concentration of the airborne 2-chloroethanol in mg/m^3 referring to $20\,°\text{C}$ and 1013 hPa
σ Concentration by volume of 2-chloroethanol in mL/m^3

8 Reliability of the method

8.1 Precision

For the determination of the standard deviation, volumes of each 5 µl of a solution containing 2.4 mg/mL of 2-chloroethanol were transferred into 10 adsorption tubes by means of a microliter syringe. After this procedure 25 L air were drawn through the adsorption tubes.

For the complete measuring method at a concentration of 480 µg/m³ (12 µg for each adsorption tube) the relative standard deviation was 3.7% and the mean variation was 8%.

8.2 Recovery rate

For the determination of the recovery rate 10 adsorption tubes were loaded with 5 µL of a solution containing 2.4 mg/mL 2-chloroethanol and a volume of 25 L air was drawn through the tubes. These tubes were analysed as described in Sect. 3.

These tests yielded an average recovery rate of >95%.
The recovery rate was not influenced by the relative humidity up to a value of 75%.

8.3 Quantification limit

The absolute quantification limit is 1.5 ng 2-chloroethanol and the relative quantification limit is 0.06 mg/m^3 for a 25 L sample of air, 1 mL sample solution and 1 µL injection volume.

8.4 Selectivity

The selectivity has to be checked in each individual case.

9 Discussion

Personal as well as static measurements can be performed by the application of this method. The method gives a time-weighted-average result of the concentration.

10 References

[1] *Deutsche Forschungsgemeinschaft* (1998) List of MAK and BAT values 1998. Maximum concentrations and biological tolerance values at the workplace. Report No 34 by the Commission for the Investigation of Health Hazards of Chemical Compounds in the work area. WILEY-VCH Verlag GmbH, Weinheim.

Authors: *J. Oldeweme, W. Merz*
Examiners: *A. Kettrup, H. Weber*

Ethylene oxide

Method number 1

Application Air analysis

Analytical principle Gas chromatography

Completed in April 1993

Summary

A measured air volume is drawn through a charcoal tube with a sampling pump. The adsorbed ethylene oxide is desorbed with a mixture of toluene and carbon disulfide, converted with hydrobromic acid to bromoethanol and determined by means of gas chromatography.

The quantitative evaluation occurs from a calibration curve whereas the determined bromoethanol peak areas are plotted against the corresponding ethylene oxide weights.

Precision: Standard deviation (rel.) $s = 5\%$
Mean variation $u = 8\%$
at a concentration of 1.1 mg/m^3
and $n = 10$ determinations

Quantification limit: 0.1 mg ethylene oxide per m^3 air
(referring to a sample volume of 12 L)

Recovery rate: $\eta = 0.95$ (95%)

Recommended sampling time: 5 h

Recommended sample volume: 12 L

Ethylene oxide

$$H_2C\!-\!CH_2$$
$$\diagdown O \diagup$$

Ethylene oxide (oxirane) is a neutral reacting colourless liquid of a sweet smell in its pure form. The molecular weight is 44.05 g/mole, the boiling point is 11.7 °C. It is miscible with water and many organic solvents.

Ethylene oxide is used in the production of ethylene glycol. The addition of ethylene oxide to alcohols, phenols and acids is a help in the production of a great number of surfactants, detergent base materials, textile auxiliaries and plastics. Additionally, ethylene oxide is used in the sterilization of medical instruments. In the food sector it serves for the disinfection of spices due to its alkylating effect.

Ethylene oxide is irritant to the skin and to the mucuous membranes. The mixtures with air are explosive in the range of 2.6–100% by volume.

It is listed in the MAK list in the Group III 2 [1]. The currently valid TRK value (1998) is 2 mg/m^3 and 1 mL/m^3 [2].

Author: *J. Oldeweme*
Examiner: *H. Muffler*

Ethylene oxide

Method number	1
Application	Air analysis
Analytical principle	Gas chromatography
Completed in	April 1993

Contents

1 General principles

A measured air volume is drawn through a charcoal tube with a sampling pump. The adsorbed ethylene oxide is desorbed with a mixture of toluene and carbon disulfide, converted with hydrobromic acid to bromoethanol and determined by means of gas chromatography (cf. Fig. 1).

The quantitative evaluation occurs from a calibration curve whereas the determined bromoethanol peak areas are plotted against the corresponding ethylene oxide weights.

2 Equipment, chemicals and solutions

2.1 Equipment

Gas chromatograph equipped with an electron capture detector
Recorder or integrator
Activated charcoal tubes, 600 mg (400/200 mg) from MSA, Pittsburgh
Suction pump, capable of operating in the range about 0.75–4.5 L/h with volume indi-
  cator or gasmeter
Mechanical flatbed shaker, about 150 cycles/min
Thermometer
Barometer
10 mL and 100 mL Volumetric flask
Glass cutter
1 mL Dosing pipette
Glass capillaries for adding the hydrobromic acid
Sample vials with polytetrafluoroethylene (PTFE) coated septum and aluminum crimp
  caps

2.2 Chemicals

Carbon disulfide, analytical grade
Toluene, analytical grade
Ethylene oxide, >99% purity
Hydrobromic acid, analytical grade, 47% (V:V)
Potassium carbonate, analytical grade

2.3 Solutions

Toluene/carbon disulfide mixture:
Mixture of toluene and carbon disulfide in a volume ratio of 99:1.

2.4 Calibration standards

Ethylene oxide stock solution:
10 mL of the desorption agent are transferred into a crimp-top vial, closed with a septum and weighed. After that 10 mL ethylene oxide are injected into the liquid with a

gas-tight syringe with roughly shaking. The solution is shaken for additional 15 min and weighed again. The difference should be about 18 mg. The solution contains about 1.8 mg/mL ethylene oxide.

Stock solution:
200 µL ($\hat{=}$ 353 mg) 2-bromoethanol are transferred into a 100 mL volumetric flask and filled up to the mark with the toluene/carbon disulfide mixture with occasionally shaking. The fresh calibration standards which contain 2–35 mg/L bromoethanol are prepared daily from the stock solution by dilution with the toluene/carbon disulfide mixture.

At a sample volume of 12 L a concentration range of 0.155–3.1 mg/m^3 ethylene oxide is detected by use of these solutions (cf. Tab. 1).

Table 1. Pipetting scheme for the preparation of calibration standards.

Volume of the stock solution µg	Final volume of the calibration standards mL	Concentration of bromoethanol mg/L	Concentration of ethylene oxide mg/m^3
5	10	1.76	0.155
10	10	3.53	0.31
50	10	17.6	1.55
100	10	35.3	3.1

3 Sample collection and preparation

Using a flow stabilized pump or a pump which is controlled by a gasmeter, the sample air is drawn through the absorbent tube at a flow rate of 2.4 L/h. The flow rate may not exceed this value. The maximum air sample volume is 12 L.

The decisive parameters for the determination of the concentration like sample volume, temperature in the gasmeter and ambient pressure have to be determined at the measuring location.

The loaded tubes can be stored only a short time without losses. To make sure, the tubes should be closed with plastic caps before they are stored.

For the desorption the contents of a loaded activated charcoal tube are transferred into a sample vial, 5 mL toluene/carbon disulfide mixture added for the ethylene oxide desorption and the vial shaken occasionally. After a desorption time of 1 hour, the liquid phase is separated, transferred into a vial, 25 µL hydrobromic acid added and the vial shaken on a flatbed shaker for 30 min. After this procedure the hydrobromic acid which was added as an excess is neutralized with about 10 mg potassium carbonate. The solution above is then transferred into a sample vial and the bromoethanol formed by the conversion of ethylene oxide with hydrobromic acid is determined by means of gas chromatography.

The conversion reaction may not be started in the presence of activated charcoal. Otherwise the formation of bromoethanol does not occur quantitatively.

4 Operating conditions for gas chromatography

Column:	Material:	Capillary column
	Length:	25 m
	Internal diameter:	0.25 mm
Stationary phase:	FFAP (polyethylene glycol)	
	Film thickness:	1 μm
Detector:	Electron capture detector	
Temperatures:	Injector:	210 °C
	Detector:	250 °C
	Oven:	Start at 80 °C
		with 10 °C/min up to 180 °C
		5 min at 180 °C
Carrier gas:	Helium 4.6	
Injection:	Splitless	
Injection volume:	1 μL	

5 Analytical determination

Replicate volumes of 1 μL each of the sample solution (cf. Sect. 3) are injected into the injector block of the gas chromatograph and the peak areas of bromoethanol are determined. Analysis and calibration have to be carried out always in one operational procedure.

6 Calibration

Volumes of 1 μL each of the calibration solutions (cf. Sect. 2.4) are injected into the gas chromatograph. The calibration curve is obtained by plotting the peak areas against the corresponding ethylene oxide weights in the elution volume (mg/5 mL toluene/carbon disulfide).

7 Calculation of the analytical result

By use of the peak area the corresponding concentration of ethylene oxide is taken from the calibration curve.
The corresponding weight concentration ρ (mg ethylene oxide/m^3 ambient air) is given by the following equation:

$$\rho = \frac{X}{V_Z \cdot \eta} \cdot \frac{273 + t_g}{273 + t_a}$$

At 20 °C and 1013 hPa:

$$\rho_0 = \rho \, \frac{273 + t_a}{293} \cdot \frac{1013 \, \text{hPa}}{p_a}$$

The corresponding volume concentration (independent of the pressure and temperature) is:

$$\sigma = \rho_0 \, \frac{24.1 \, \text{L/mole}}{44 \, \text{g/mole}}$$

$$\sigma = \rho \cdot \frac{273 + t_a}{p_a} \cdot \frac{1013 \, \text{hPa}}{293} \cdot \frac{24.1 \, \text{L/mole}}{44 \, \text{g/mole}}$$

$$\sigma = \rho \cdot \frac{273 + t_a}{p_a} \, 1.89 \, \frac{\text{hPa} \cdot \text{mL}}{\text{mg}}$$

At $t_a = 20$ °C and $p_a = 1013$ hPa:

$$\sigma = \rho \cdot 0.548 \, \frac{\text{mL}}{\text{mg}}$$

Legend:

X Weight of ethylene oxide in mg in the sample solution
V_Z Sample volume in m^3
η Recovery rate
t_g Temperature in the gas flowmeter in °C
t_a Temperature in the ambient air in °C
p_a Ambient pressure in hPa
ρ Concentration by weight of airborne ethylene oxide in mg/m^3 referring to t_a and p_a (see above)
ρ_0 Concentration of the airborne ethylene oxide in mg/m^3 referring to 20 °C and 1013 hPa
σ Concentration by volume of ethylene oxide in mL/m^3

8 Reliability of the method

8.1 Precision

By use of a gas-tight syringe 7 µL ($\hat{=}$ 13 µg) ethylene oxide are transferred each in 10 adsorption tubes to determine the standard deviation. After that 12 L air was drawn through each tube. For the complete measuring method at a concentration of 1096 µg/m^3 (13 µg for each adsorption tube) the relative standard deviation was 5% and the mean variation was 8%. For the further determination of the precision volumes of 2, 5, 10, 15, 25 and 35 µL of the ethylene oxide stock solution (cf. Sect. 2.4) were injected six-fold into a glass tube. For five hours laboratory air was drawn through the glass tube and two serially connected adsorption tubes at a flow rate of 2.4 L/h. At 12 L air volume the injected ethylene oxide quantities corresponded with concentrations between 0.2 and 5 mg/m^3. The obtained standard deviations ranged between 6 and 12%.

8.2 Recovery rate

At a sample volume of 12 L and a flow rate of 2.4 L/h the recovery rates were over 95% under the conditions as described in Sect. 8.1.

8.3 Quantification limit

The absolute quantification limit is 0.25 ng ethylene oxide. The relative quantification limit is 0.1 mg/m^3 $\hat{=}$ 0.05 mL/m^3 ethylene oxide for 12 L air, 5 mL sample solution and 1 µL injection volume.

8.4 Sources of error

Under the experimental conditions for the sampling, calibration and analysis as mentioned no interferences are expected.

9 Discussion

Personal as well as static measurements can be performed by the application of this method. A selective measurement is possible due to the derivatization of the ethylene oxide and the application of an electron capture detector. For the gas chromatographic determination a quartz capillary column can be used as well as a packed column. The method gives a time-weighted-average result of the concentration. Simultaneously to the method described here methods in which the derivatization of ethylene oxide into bromoethanol is performed during the sampling were developed [3, 4].

10 References

[1] *Deutsche Forschungsgemeinschaft* (1998) List of MAK and BAT values 1998. Maximum concentrations and biological tolerance values at the workplace. Report No 34 by the Commission for the Investigation of Health Hazards of Chemical Compounds in the work area. Wiley-VCH Verlag GmbH, Weinheim.

[2] *Bundesministerium für Arbeit und Sozialordnung* (1998) TRGS 900: Grenzwerte in der Luft am Arbeitsplatz – Luftgrenzwerte. Technische Regeln und Richtlinien des BMA zur Verordnung über gefährliche Stoffe. BArbBl. 10/1996, p. 106–128, last supplement 10/1998, p. 73–74.

[3] *Cummins KJ, Schultz GR, Lee JS, Nelson JH, Reading JC* (1987) The development and evaluation of a hydrobromicacid-coated sampling tube for measuring occupational exposures to ethylene oxide. Am Ind Hyg Assoc J 48 (6): 563–573.

[4] *Lefevre C, Ferrari P, Delcourt J, Guenieret JP, Muller J* (1986) Ethylene oxide pollution evaluation. Part III: Sampling on HBr treated charcoal tubes. Chromatographia 21 (5): 269–273.

Author: *J. Oldeweme*
Examiner: *H. Muffler*

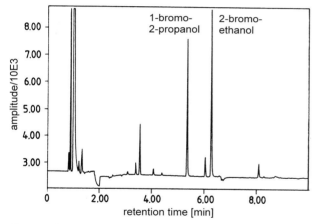

Fig. 1. Chromatogram of the determination of ethylene oxide and propylene oxide.

Halogenated narcosis gases (Halothane, Enflurane, Isoflurane)

Method number   1

Application   Air analysis

Analytical principle Gas chromatography

Completed in   November 1994

Summary

The method permits the simultaneous determination of the common narcosis gases halothane (2-bromo-2-chloro-1,1,1-trifluoroethane), enflurane (2-chloro-1,1,2-trifluoro-
ethyl difluoromethyl ether) and isofluorane (1-chloro-2,2,2-trifluoroethyl difluoromethyl
ether) in the workplace air. Samples can be collected by active sampling (with sampling
pumps) or by passive (diffusion) sampling.
With a sampling pump measured air volumes are drawn through sampling tubes containing XAD-4. The halogenated narcosis gases are adsorbed. In diffusion sampling,
contact between adsorbent and narcosis gases occurs by diffusion. The adsorbed halogenated narcosis gases are desorbed thermally and then analysed in a gas chromatograph equipped with a flame ionization detector. Calibration standards of known composition are used for the quantitative evaluation. The peak areas are linearly dependent
on the concentrations of halothane, enflurane and isoflurane.

Precision: *Halothane*: (active sampling and diffusion sampling)
                        Standard deviation (rel.) $s = 0.37\%$ and 3.6%
                        Mean variation $u = 1.0\%$ and 9.9%
                        for $n = 5$ determinations and $c = 75$ mg/m^3
                        Enflurane: (active sampling and diffusion sampling)
                        Standard deviation (rel.) $s = 0.81\%$ and 4.2%
                        Mean variation $u = 2.2\%$ and 11.7%
                        for $n = 5$ determinations and $c = 50$ mg/m^3
                        Isoflurane: (active sampling and diffusion sampling)
                        Standard deviation (rel.) $s = 3.0\%$ and 3.2%
                        Mean variation $u = 8.5\%$ and 9.0%
                        for $n = 5$ determinations and $c = 25$ mg/m^3

Quantification limit:	1 mg/m^3 (active sampling) and 1–2 mg/m^3 (diffusion sampling)
Recovery rate:	$\eta > 0.98$ (> 98%)
Recommended sampling volume and time:	about 200 mL at a flow rate of 1–4 mL/min by active sampling 4–8 hours by diffusion sampling

Halogenated narcosis gases

For narcosis of patients in medical operations the volatile anaesthetics halothane, enflurane and isoflurane are used together with laughing gas.

Halothane (2-bromo-2-chloro-1,1,1-trifluoroethane)

$F_3C-CHBrCl$

Halothane is a well-accepted anaesthetic, also known under the synonyms fluothane and rhodialothan. The MAK value has been 5 mL/m^3 (41 mg/m^3) since 1978 [1]. Maximum levels of exposure are restricted by the peak limitation value of twice the MAK value (average value during 30 minutes). It is said to have prenatal toxic effects.

Enflurane (2-chloro-1,1,2-trifluoroethyl difluoromethyl ether)

CHF_2-O-CF_2CHClF

Enflurane (synonym: ethrane) is a stable non-flammable inhalation anaesthetic. Since 1994 the MAK value for enflurane has been established at 20 mL/m^3 (150 mg/m^3) [1].

Isofluorane (1-chloro-2,2,2-trifluoroethyl difluoromethyl ether)

$CF_3-CHCl-O-CHF_2$

Isoflurane is also an inhalation anaesthetic, also known under the synonyms forane, forene and aerrane.

Author: *M. Tschickardt*
Examiners: *E. Flammenkamp, C. Madl*

Halogenated narcosis gases (Halothane, Enflurane, Isoflurane)

Method number   1

Application   Air analysis

Analytical principle Gas chromatography

Completed in   November 1994

Contents

1 General principles

The method permits the simultaneous determination of the common narcosis gases halothane (2-bromo-2-chloro-1,1,1-trifluoroethane), enflurane (2-chloro-1,1,2-trifluoro-
ethyl difluoromethyl ether) and isofluorane (1-chloro-2,2,2-trifluoroethyl difluoromethyl
ether) in the workplace air. The samples can be collected by active sampling (with
sampling pumps) or by passive (diffusion) sampling.
With active sampling measured air volumes are drawn through an adsorption tube containing the adsorbent XAD-4 to collect the narcosis gases (cf. Fig. 1). In diffusion sampling, contact between adsorbant and narcosis gases occurs by diffusion.
With the method described here the sampling tube recommended for active sampling
can also be used for passive sampling. In both cases the analysis is carried out after
thermal desorption and enrichment of the sample in a packed cold trap with a gas chromatograph equipped with an FID (cf. Fig. 2).
Calibration standards of known composition are used for the quantitative evaluation.
The peak areas are linearly dependent on the concentrations of halothane, enflurane
and isoflurane.

2 Equipment and chemicals

2.1 Equipment

Adsorption tubes of stainless steel, 6.3 mm x 90 mm, 5 mm internal diameter
Sampling pump, flow rate 1–4 mL/min (e. g. SKC type 222-1-4, MTC, Müllheim)
Thermometer
Barometer
Diffusion caps (diffusion sampling) (e. g. from Perkin-Elmer, order number 126 433)
Gas chromatograph equipped with a thermal desorber (e. g. ATD 400 from Perkin-El-
    mer) and a flame ionization detector
Computerized data collection and integration system
Apparatus for dynamic calibration or dynamic test gas apparatus
Caps to seal the adsorption tubes (e. g. Swagelok®, Teflon or aluminum)

2.2 Chemicals

Halothane 99.9% purity or calibration gas mixture
Enflurane 99.9% purity or calibration gas mixture
Isoflurane 99.9% purity or calibration gas mixture
Helium (carrier gas) 99.996% purity
Nitrogen 99.999% purity

XAD-4 adsorber resin, 0.2–0.4 mm, analytical grade (e.g. Serdolit AD-4 from Serva, Heidelberg, order number 42 436)
Tenax TA, 60–80 mesh (e.g. from Chrompack, Frankfurt)

2.3 Pretreatment of the adsorption tubes

15 g of the adsorber resin is purified in a suitable stainless steel tube in a stream of nitrogen or helium at a flow rate of 100 mL/min at 150 °C for 16 hours. This amount of adsorbent is sufficient to pack about 30 adsorption tubes. Each adsorption tube is packed with 450 mg of XAD-4. The adsorbent is fixed between two stainless steel sieves. One of the sieves is placed exactly 15 mm from the end of the tube.
Before use the adsorption tubes are heated at 150 °C in the thermal desorber and the blank values checked.
They are closed with suitable caps for storage.

3 Sample collection

Sample collection can be carried out by local monitoring or by personal air sampling. The parameters which are important for the determination of the concentration such as sample volume, temperature, atmospheric pressure and relative humidity are noted in a sampling protocol as well as the times at the beginning and the end of the sample collection.
The samples are collected in the breathing area. The opening of the adsorption tube should not be obstructed.

3.1 Active sampling

Using a sampling pump the air to be sampled is drawn continuously through the adsorption tube at a flow rate of 1–4 mL/min. After the sampling is finished the loaded adsorption tube is closed with caps at both ends. A minimum sample volume of 200 mL air is recommended.
The method has been tested and may be used in the range of 5–80% relative humidity.

3.2 Diffusion sampling

Before sample collection the cap on the end of the tube intended for diffusion sampling is replaced by a diffusion cap. A sampling time of 4–8 hours is recommended.

4 Analytical operating conditions

4.1 Thermal desorption

The adsorption tubes are put into a thermal desorber, heated and the adsorbed components transferred into a cold trap by a carrier gas. When desorption is complete, the
cold trap is heated so that the substance reaches the GC column as a concentrated
band.
The instrumental parameters for the ATD 400 gas chromatograph are as follows:

Desorption temperature:	150 °C
Desorption time:	5 min
Transfer tubing:	100 °C
Cold trap (adsorption):	−30 °C
Cold trap (injection):	300 °C
Weight of the adsorbent in the cold trap:	20 mg Tenax TA (60–80 mesh)
Carrier gas:	Helium
Input split:	41 mL/min
Desorb flow:	10 mL/min
Output split:	28 mL/min

The instrumental conditions have to be modified for other types of thermal desorbers.

4.2 Operating conditions for gas chromatography

Column:	Length:	30 m
	Internal diameter:	0.25 mm
Stationary phase:	DB-Wax	
	Film thickness:	0.5 µm
Detector:	Flame ionization detector	
Temperatures:	Oven:	10 min 50 °C isothermal, then 8 °C/min to 120 °C, hold for 1 min
	Detector:	200 °C
Carrier gas:	Helium:	1.25 kPa (1.7 mL/min)

5 Analytical determination and calibration

100 mL calibration gas taken from a self-prepared calibration atmosphere is drawn
through adsorption tubes in order to check the analytical procedure. The concentration

range of the calibration gases should be between 10 and 500 % of the limit values (e. g. calibration gas preparation as described in [2]). The parameters during the adsorption of the calibration gas (pressure, temperature) are recorded.

After preparation of the thermal desorber and the gas chromatograph (cf. Sect. 4.1 and 4.2) the calibration samples and the workplace samples are analysed. The peak areas are plotted against the concentrations of calibration samples used.

6 Calculation of the analytical result

The concentration in the sample is calculated from the peak areas of the calibration samples and the workplace samples according to the following equation:

$$\rho = \frac{A_p \cdot \rho_k \cdot V_k}{A_k \cdot V_p}$$

Legend:

ρ Concentration of the analyte in the air sample (mg/m^3)
A_p Peak area of the analyte in the sample
A_k Peak area of the analyte in the calibration gas
ρ_k Concentration of the analyte in the calibration gas (mg/m^3)
V_k Sampled volume of the calibration gas (mL)
V_p Sampled volume of the sample (mL); in diffusion sampling: sampling time (min) x uptake rate (mL/min) (cf. Tab. 1)

Before inserting into the equation given above, the volumes have to be corrected to standard conditions.

7 Reliability of the method

7.1 Precision

In order to determine the precision of the method, calibration gas concentrations of 75 mg/m^3 halothane, 50 mg/m^3 enflurane and 25 mg/m^3 isoflurane were generated with a dynamic calibration gas apparatus and five samples collected by active sampling and five by diffusion sampling. Then the samples were processed and analysed. The following data were obtained:

		Precision	
		Active sampling	Diffusion sampling
Halothane:	Standard deviation (rel.)	$s = 0.37\%$	$s = 3.6\%$
	Mean variation	$u = 1.0\%$	$u = 9.9\%$
Enflurane:	Standard deviation (rel.)	$s = 0.81\%$	$s = 4.2\%$
	Mean variation	$u = 2.2\%$	$u = 11.7\%$
Isoflurane:	Standard deviation (rel.)	$s = 3.0\%$	$s = 3.2\%$
	Mean variation	$u = 8.5\%$	$u = 9.0\%$

7.2 Recovery rate

The recovery rate was determined by multiple experiments with desorption of loaded tubes. The concentrations were between 4 and 200 mg/m^3 for halothane and 7 and 400 mg/m^3 for the other two narcosis gases. The desorption rate for all 3 components was greater than 98% for both active and diffusion sampling.

7.3 Quantification limit

With active sampling, the detection limit is 1 mg/m^3 for each component for a sample volume of 200 mL. In diffusion sampling the quantification limit is 1–2 mg/m^3 for a sampling time of 4–8 hours.

7.4 Specificity

The specificity of the method depends especially on the type of GC column used. In practice, the column mentioned above has proved suitable. Iso-propanol and ethanol, which are present in the air of operating theatres, are separated from the anaesthetics. If interfering substances are present another separation phase has to be selected.

8 Discussion

Although the method was tested with XAD-4 tubes, other adsorbents such as Chromo-sorb 106 can also be used. If other adsorbents or tubes of different dimensions are used, the break-through volumes and the analytical parameters have to be checked (cf. Tab. 2). The suitability of the adsorbent with respect to blank value and desorption rate has to be checked for each new batch.

The adsorption tubes must be heated in the desorber immediately before sampling is performed because after storage for longer periods, interfering substances released from the cap gaskets can diffuse into the tube and be adsorbed on the collection phase.

The water vapour in humid air (about 60% relative humidity) hardly displaces adsorbed substances from the surface of the adsorbant.

In the temperature range of 20–25 °C the tubes could be stored for one week after sampling without loss.

Apparatus: Thermal desorber ATD 400 and gas chromatograph 8700 of Perkin-Elmer.

9 References

[1] *Deutsche Forschungsgemeinschaft* (1998) List of MAK and BAT values 1998. Maximum concentrations and biological tolerance values at the workplace. Report No 34 by the Commission for the Investigation of Health Hazards of Chemical Compounds in the work area. WILEY-VCH Verlag GmbH, Weinheim.
[2] *Verein Deutscher Ingenieure (VDI)* (1981) VDI-Richtlinie 3490 Blatt 8, Prüfgase-Herstellung durch kontinuierliche Injektion. Beuth Verlag, Berlin.
[3] *Perkin Elmer* (1991) The selection of adsorbents for diffusive sampling. Thermal Desorption Applications No 11, Überlingen.

Author: *M. Tschickardt*
Examiners: *E. Flammenkamp, C. Madl*

Table 1. Uptake rate (mL/min) of diffusion sampling on XAD-4 (60–80 mesh) with stainless steel tube for type ATD from Perkin-Elmer.

Sampling time min	Isoflurane	Halothane	Enflurane
65	0.550	0.560	0.590
120	0.590	0.560	0.500
180	0.525	0.540	0.545
240	0.610	0.580	0.580
Mean	0.569	0.560	0.554
Standard deviation	0.038	0.016	0.041

Table 2. Break-through volumes and retention volumes for 420 mg of the adsorbent XAD-4 (60–80 mesh) in each adsorption tube.

	Isoflurane	Halothane	Enflurane
Break-through volume * (mL)	20 000	16 000	27 000
Retention volume *	212 000	169 000	238 000

* Break-through volume (in mL/420 mg XAD-4 at 30 °C) and retention volume (in L/g XAD-4 at 20 °C) [3]

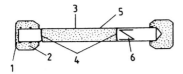

1 Cap or
2 Diffusion cap
3 Adsorbent
4 Stainless steel sieves
5 Adsorption tube
6 Spring

Fig. 1. Sampling tube type ATD.

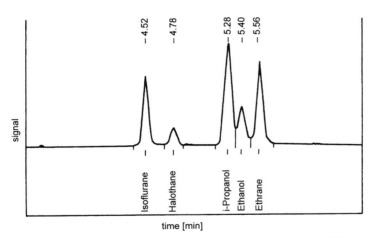

Fig. 2. Separation of the halogenated anaesthetics from solvents (GC parameters cf. Sect. 4).

Organotin compounds
(Species analysis)

Method number   1

Application   Air analysis

Analytical principle Gas chromatography

Completed in   October 1992

Summary

By means of a sampling pump measured air volumes are drawn from the breathing area through adsorption tubes which are filled with the protonated type of a wet cation exchange resin. Solid particles and aerosols are collected on a preconnected glass fibre filter. The gaseous mono-, di- and trialkyltin compounds are bound ionically and tetra-alkyltin is bound adsorptively to the ion exchange resin. After desorption with acidified diethyl ether the mono-, di- and trialkyltin chlorides are reacted with pentylmagnesium bromide to the corresponding tetraalkylstannanes (RPe_3Sn, R_2Pe_2Sn, R_3PeSn). They are transferred to a silica gel column for clean-up and eluted with hexane. The polar compounds remain on the column. The tetraalkylstannanes are separated and determined by means of capillary gas chromatography. If interferring components are present (bad resolution of the peaks in the chromatogram, difficulties in the assignments of peaks) they can be identified and quantified by means of a mass selective detector. Dihexyltin dichloride is used as an internal standard.

Precision:	Standard deviation (rel.) $s = 1.4$ and 8.6%
	Mean variation $u = 2.9$ and 19.4%
	at concentrations of about 0.15 and 0.015 mg tin per m^3 air and $n = 10$ determinations (measured for 4 different organotin compounds)
Detection limit:	0.1 µg tin per m^3 air for each individual component (at a sample volume of 100 L)
Recovery rate:	$\eta = 0.56-0.94$ (56–94%) (cf. Tab. 1)
Recommended sampling time:	1 h
Recommended sample volume:	360 L

Organotin compounds

Tin compounds with at least one tin-carbon bond in the molecule are named organotin compounds:

R_4Sn	Tetraorganotin compounds
R_3SnX	Triorganotin compounds
R_2SnX_2	Diorganotin compounds
$RSnX_3$	Monoorganotin compounds

Legend:

R = Alkyl-, cycloalkyl-, carbobutoxyethyl-, acrylic hydrocarbon substituents
X = Anionic groups like halogene, $-OH$, $-OR'$, $-SR'$, $-OOCR'$, $-NR_2'$

The tetraorganotin compounds mainly serve as intermediates for the production of mono-, di- and triorganotin compounds. Triorganotin compounds are used as biocides against mites, fungi, bacteria and algae. The most important fields of application for tributyltin are wood preservatives, textile preservatives, biocide treatment of sealing materials and as antifouling paints for ships. Triphenyltin and tricyclohexyltin are applied as plant protection agents. Di- and monoorganotin compounds (especially methyltin, butyltin and octyltin) are used as PVC stabilizers, as catalysts for polymerizations, esterifications and trans-esterifications. Methyltin chloride is applied in the refinement of glass surfaces.
The MAK value of organotin compounds is 0.1 mg/m^3 (referring to tin) [1]. The symbol "H" in the MAK list indicates the risk of skin absorption. The tri-n-butyltin compounds are an exception. Here the MAK is 0.05 mg/m^3 (referring to TBTO) [1]. This corresponds with a concentration of 0.02 mg/m^3 referring to tin.
The toxicology of the organotin compounds is described in detail in the DFG method "Tin in urine" [2] and in "The evaluation of occupational toxicants" by the DFG [3].

Authors: *H.A. Müller, U. Schings-Hartmann*
Examiners: *W. Kleiböhmer, K. Cammann*

Organotin compounds
(Species analysis)

Method number   1

Application   Air analysis

Analytical principle Gas chromatography

Completed in   October 1992

Contents

1 General principles

By means of a sampling pump measured air volumes are drawn from the breathing area through adsorption tubes which are filled with the protonated type of a wet cation exchange resin. Solid particles and aerosols are collected on a preconnected glass fibre filter. The gaseous mono-, di- and trialkyltin compounds are bound ionically and tetra-alkyltin is bound adsorptively to the ion exchange resin. After desorption with acidified diethyl ether the mono-, di- and trialkyltin chlorides are reacted with pentylmagnesium bromide to the corresponding tetraalkylstannanes (RPe_3Sn, R_2Pe_2Sn, R_3PeSn). They are transferred to a silica gel column for clean-up and eluted with hexane. The polar compounds remain on the column. The tetraalkylstannanes are separated and determined by means of capillary gas chromatography. If interferring components are present (bad resolution of the peaks in the chromatogram, difficulties in the assignments of peaks) they can be identified and quantified by means of a mass selective detector. Dihexyltin dichloride is used as an internal standard.

2 Equipment, chemicals and solutions

2.1 Equipment

Capillary gas chromatograph equipped with a flame ionization detector or a mass selective detector, on-column injector, option for the recording and evaluation of chromatograms (e. g. recorder, integrator or work station).

Adsorption tubes: Eppendorf pipette tip (blue) filled with about 0.5 g wet cation exchange resin placed between two stoppers of silanized glass wool inserted in a suction finger (glass tube melted at one end, length about 10 cm, internal diameter about 1 cm, equipped with olive adapter and screw closure with hole (cf. Fig. 1).

Glass fiber filter, diameter 13 mm, thickness 550 µm, pore size about 50–60 µm (e.g. Sartorius SM 6, # S 13400)

Suction pump, pumping capacity about 6 L/min

Gasmeter

Thermometer

Barometer

Glass tube (length 15 cm, diameter 2 cm) equipped with glass frit bottom and delivery cock (chromatography column)

Solid phase extraction system (e. g. VacElut of ICT)

10 mL Medical disposable syringes with Luer tip (e. g. of Henke-Sass, Wolf GmbH)
Eppendorf pipettes, variable
Eppendorf pipette tips, blue, for 100–1000 µL pipettes
1 and 100 mL Bulb pipettes
100 mL Volumetric flasks
50 mL Ampoules
5 mL Injection vials with aluminium closure caps and PTFE coated septa
Crimper
1 µL Liquid syringe for gas chromatography
Ultrasonic bath
Analysis balance

2.2 Chemicals

Cation exchange resin Amberlite CG 120 I (Rohm & Haas, Philadelphia, USA; e. g.
  EGA Chemie), grain size 100–200 mesh, counter-ion: Na$^+$
Butyltin trichloride, 95 % e. g. from Aldrich [20, 105-7]
Dibutyltin dichloride, 97 % e. g. from Aldrich [20, 549-4]
Tributyltin chloride, 96 % e. g. from Aldrich [T 5, 020-2]
Tetrabutyltin, 96 % e. g. from Aldrich [T 600-8]
Dihexyltin dichloride as internal standard, e. g. from Acima, CH-9470 Buchs
Pentylmagnesium bromide, 2.0 M in diethyl ether, e. g. from Aldrich [29, 099-8]
2-Propanol, analytical grade, e. g. from Merck
Diethyl ether, analytical grade, e. g. from Merck
t-Butyl methyl ether, residue analysis, e. g. from Merck
n-Hexane, analytical grade
Hydrochloric acid, highest purity, $w = 30$ % (e. g. Suprapur from Merck)
Sulfuric acid 0.5 M, analytical grade
Sodium hydroxide, 2 M, analytical grade
Hydrochloric acid, 2 M (prepared from the hydrochloric acid 30 % as mentioned)
Silica gel 60, grain size 0.063–0.200 mm, 70–230 mesh, for column chromatography
  (e. g. from Merck)
Glass wool, silanized

2.3 Solutions

Diethyl ether (hydrochloric):
Add 100 mL diethyl ether and 5 mL of 30 % hydrochloric acid to a separating funnel.
After shaking the flask, the aqueous phase is separated and disposed of.

2.4 Calibration standards

The following experimental procedure is generally applied for all organotin compounds (except the estertin compounds). The preparation of the calibration standards for the butyltin compounds is described as an example.

Initial solution:
300–400 mg each of tetrabutyltin, dibutyltin dichloride and butyltin trichloride and 800–1000 mg tributyltin chloride are exactly weighed into a 100 mL volumetric flask and diluted up to the mark with *t*-butyl methyl ether.

Stock solution:
10 mL of the initial solution are pipetted into a 100 mL volumetric flask and diluted up to the mark with *t*-butyl methyl ether. This stock solution contains
102.6–136.8 mg Sn/L as tetrabutyltin
291.8–364.7 mg Sn/L as tributyltin chloride
117.2–156.2 mg Sn/L as dibutyltin dichloride
126.2–168.3 mg Sn/L as monobutyltin trichloride

Calibration standard solutions:
The calibration standard solutions are prepared from the stock solution by dilution with *t*-butyl methyl ether (Bu$_4$Sn).

Table 1. Pipetting scheme for the preparation of calibration standards.

Stock solution No	mL	Final volume mL	Concentration Bu$_4$Sn mg Sn/L	Bu$_3$SnCl mg Sn/L	Bu$_2$SnCl$_2$ mg Sn/L	BuSnCl$_3$ mg Sn/L
1	10.00	100	10.26–13.68	29.18–36.47	11.72–15.62	12.62–16.83
2	5.00	100	5.13– 6.84	14.59–18.24	5.86– 7.81	6.31– 8.42
3	2.00	100	2.05– 2.74	5.84– 7.29	2.34– 3.12	2.52– 3.27
4	1.00	100	1.03– 1.37	2.92– 3.65	1.17– 1.56	1.26– 1.68
5	0.75	100	0.77– 1.03	2.19– 2.74	0.88– 1.17	0.95– 1.26
6	0.50	100	0.52– 0.69	1.46– 1.83	0.59– 0.78	0.63– 0.84

A volume of 200 µL of the working solution of the internal standard is added to 200 µL of the prepared calibration standard solutions as mentioned before. Then 5 mL of hexane are added and the solution is alkylated and treated according to Sect. 3.6. 1 µL of each solution is analysed for the determination of the peak area correction factor referring to the internal standard.

Preparation of the solution of the internal standard:
400–500 mg dihexyltin dichloride are weighed into a 100 mL volumetric flask and diluted up to the mark with *t*-butyl methyl ether (stock solution).
1 mL of the stock solution are pipetted into a 100 mL volumetric flask and diluted up to the mark with *t*-butyl methyl ether. The solution (working solution) prepared in this way contains 13.2–16.5 mg Sn/L as dihexyltin dichloride.

3 Sample collection and preparation

3.1 Cleaning of the vessels (glass ware)

After the normal cleaning procedure of the laboratory vessels all glass ware has to be cleaned with acidified 2-propanol until no more tin is detectable in the wash solution. The glass ware is dried at air.

3.2 Preparation of the exchange resin

The Na^+ type of the exchange resin is allowed to swell in deionized water for 24 hours. The turbid water phase (due to the content of fine grained resin) is decanted and the resin is washed until the water is clear. The resin is filled into a chromatographic column up to a height of about 5 cm, washed slowly with 50 mL NaOH (2 M) and then with deionized water until the eluate is neutral. The resin is converted into the H^+ type with 50 mL HCl (2 M) and eluted with deionized water until chloride is not detectable in the eluate. The resin prepared in this way is stored under water. Before use it has to be washed again with water.

3.3 Preparation of the adsorption unit

A small piece of silanized glass wool is placed in the conical end of an Eppendorf pipette tip (blue) and the treated exchange resin added to a height of 25 mm. It is closed with silanized glass wool and first washed with 1 mL acidified 2-propanol (5 mL 30% hydrochloric acid in 100 mL 2-propanol) and then with 1 mL of deionized water. The prepared pipette tip is positioned into the glass vessel and sealed with aluminium foil until use.

3.4 Sampling and sample storage

For one hour and at a flow rate of 6 L/min air is drawn through an adsorption tube containing exchange resin by means of a flow-stabilized pump or a sampling device controlled by a gasmeter. A glass fibre filter is mounted before the inlet of the adsorption tube. If necessary two exchanger tubes have to be connected in series. The decisive parameters for the concentration determination like sample volume, temperature in the gasmeter as well as ambient pressure and ambient temperature at the measuring location have to be determined.

Loaded adsorption tubes can be stored at least for one week without loss. The tubes should be closed with plastic caps (free of tin) and kept in the dark (if possible in a refrigerator).

3.5 Preparation and conditioning of the silica gel cartridges

Silica gel is doped with 5% deionized water and allowed to swell for 24 h. 10 mL disposable syringes are 3/4 filled with the conditioned silica gel.

3.6 Desorption, derivatization and clean-up

After sampling the exchange resin and the glass fibre filter are transferred into a 50 mL vial and 200 μL of internal standard solution are pipetted into the vial. Desorption is effected by treatment with 2 mL acidified diethyl ether in an ultrasonic bath for 2 min. 5 mL n-hexane is added and the solution is alkylated dropwise with 1 mL pentylmagnesium bromide solution in an ice bath. After the reaction has subsided, a further 1 mL pentylmagnesium bromide solution is used to complete the reaction for another 5 min in the ultrasonic bath. In an ice bath 0.5 M sulfuric acid (2–3 mL) is dropped into the solution until two clear, colourless phases are formed. The organic phase is transferred into a 50 mL vial using an Eppendorf pipette and the solvent is evaporated carefully in a nitrogen flow to a residual volume of 200 μL. Thus removes the diethyl ether which interferes with the following clean-up with silica gel. 1 mL n-hexane is added to the residue and the clean-up is carried out by use of a conditioned silica gel cartridge. It is desorbed with 15 mL hexane and the eluate is concentrated to 200 μL in a nitrogen flow. Losses of organotin species may occur during the concentrating process.
1 μL of this solution is injected into the gas chromatograph.

3.7 Determination of the blank value

An unloaded adsorption tube is treated as described in Sect. 3.6. If tetraalkylstannanes are found in the chromatogram (and the characteristic isotopic pattern of tin in the mass spectrum) either the source of pollution has to be detected and removed (e. g. traces of tributyl-pentyl-tin were sometimes found in the pentylmagnesiumbromide) or the blank value has to be considered in the analysis.

4 Operating conditions for capillary gas chromatography equipped with flame ionization detector or mass selective detector

4.1 Operating conditions for gas chromatography

Column:	Fused silica capillary column
	Length: 12 m
	Internal diameter: 0.31 mm
Stationary phase:	Methylsilicone rubber
	Film thickness: 1.05 μm

Precolumn:	Fused silica capillary column
	Length: 1 m
	Internal diameter: 0.53 mm
	desactivated, uncoated
Detector:	Flame ionization detector or
	mass selective detector
Temperatures:	Column: 60 °C 1 min isothermal then
	with 15 °C/minute up to 280 °C
	Flame ionization detector: 300 °C
	Transfer-line from the ion source
	to the mass spectrometer: 280 °C
Carrier gas:	Helium 5.0 (99.999 % by volume helium)
	Column head pressure: 1020 hPa
	Column flow: 1 mL/min
Sample injection:	On column
Injection volume:	1 μL

An example of a gas chromatogram is represented in Fig. 3.

4.2 Operating conditions for the mass selective detector

Both detection modes SCAN mode and Multiple Ion Detection (MID) are necessary for detection by the mass selective detector. The ionization energy is 70 eV in both cases. First a total ion current chromatogram of the alkylated and prepared calibration solutions (cf. Sect. 2.4 and 3.6) is recorded in the SCAN mode. Significant masses and the ratios of intensities of two ions or ion groups for each tin compound are taken from the obtained mass spectra. (Normally for tetraalkylstannanes the significant ions of the isotopic clusters R_3Sn^+ or $R_2R'Sn^+$ are preferred because they are very intense and interferences are not expected). The identification and quantification of the alkyltin compounds in the sample occurs by use of the MID technique based on the retention time, the mass number and the correct relationship mass number/intensity.

The following parameters for the SCAN mode and the MID mode were selected for the data acquisition:

SCAN
Mass range: 100–550 amu
Rate: 0.82 scans/s

MID

Component	Significant ions m/z	Qualifier ion m/z
Bu_4Sn	291, 289, 235, 233	289
Bu_3SnPe	305, 303, 235, 233	303
Bu_2SnPe_2	319, 317, 249, 247	317
$BuSnPe_3$	319, 317, 249, 247	317
Pe_4Sn	333, 331, 263, 261	331
Hex_2SnPe_2	347, 345, 277, 275	345

Dwell time (measuring time)/ion: 50 ms each
Electron multiplier voltage: 1800 V

5 Analytical determination

5.1 Analysis by means of a flame ionization detector

By means of a suitable GC injection syringe, 1 μL of the sample solution (prepared according to Sect. 3) are injected into the gas chromatograph and analysed as described in Sect. 4.1. To check the results, repeated analyses are performed and the peak areas of the alkyltin compounds relative to the internal standards are determined. Calibration standards are analysed with each batch of sample and also regularily for quality control.

5.2 Analysis by means of a mass selective detector

In order to define the significant masses, to determine the ratio mass number/intensity and to check the retention times of the tetraalkyltin compounds exactly 1 μL of the alkylated and prepared calibration standard solution No 4 (cf. Sect. 2.4 and 3.6) is analysed in the SCAN mode of the mass spectrometer under the operating conditions mentioned in Sect. 4.2. By use of the mass spectrometric parameter which are determined this way 1 μL of the sample solution obtained according to Sect. 3 is injected into the gas chromatohgraph and analysed in the MID mode.

To check the results, repeated analyses are performed and the peak areas of the alkyltin compounds relative to the internal standards are determined. Calibration standards are analysed with each batch of sample and also regularily for quality control.

6 Calibration

The quantitative evaluation occurs in accordance with the method of the internal standard (dihexyltin dichloride). To make the calibration curve the calibration standards
1–6 (prepared according Sect. 2.4 and alkylated according to Sect. 3.6) are analysed
by gas chromatography corresponding with the methods described in Sect. 5.1 and
5.2. The peak areas of the alkyltin compounds relative to the internals standard are
determined. The linearity of the detector was tested for the following concentrations:

Bu_4Sn:	12.0 µg–0.24 µg	Sn/mL
Bu_3SnX:	30.5 µg–0.60 µg	Sn/mL
Bu_2SnX_2:	17.5 µg–0.34 µg	Sn/mL
$BuSnX_3$:	15.0 µg–0.30 µg	Sn/mL
Pe_2SnX_2:	16.0 µg–0.32 µg	Sn/mL
Hex_2SnX_2:	5.2 µg–0.26 µg	Sn/mL

Referring to a sample volume of 100 L of air this corresponds with the range of 0.02–
1.5 fold of the currently valid MAK value for organotin.

7 Calculation of the analytical result

On the basis of the obtained peak area relationships of the tetraalkyltin compounds and
the internal standard the quantity of tin in milligram of the tetraalkyltin compound used
can be taken from the calibration curve, if the same weighed sample of the internal
standard in the calibration solutions and in the sample solutions are used.
The concentration by weight ρ (µg tin per m³ air) is calculated as follows:

$$\rho = \frac{X}{V_Z \cdot \eta} \cdot \frac{273 + t_g}{273 + t_a}$$

At 20 °C and 1013 hPa:

$$\rho_0 = \rho \, \frac{273 + t_a}{293} \cdot \frac{1013 \, \text{hPa}}{p_a}$$

The corresponding concentration by volume σ – independent of pressure and temperature – is:

$$\sigma = \rho_0 \, \frac{24.1 \, \text{L} \cdot \text{mole}^{-1}}{M} = \rho \cdot \frac{273 + t_a}{p_a} \cdot \frac{1013 \, \text{hPa}}{293} \cdot \frac{24.1 \, \text{L} \cdot \text{mole}^{-1}}{M}$$

For tin ($M = 118.7$ g mole^{-1}):

$$\sigma = \rho \cdot \frac{273 + t_a}{p_a} \, 0.702 \, \frac{\text{hPa} \cdot \text{mL}}{\text{mg}}$$

At $t_a = 20\,°C$ and $p_a = 1013$ hPa:

$$\sigma = \rho \cdot 0.203 \, \frac{\text{mL}}{\text{mg}}$$

Legend:

V_z Sample volume in m^3

η Recovery rate

t_g Temperature in the gasmeter in °C

t_a Temperature of the ambient air in °C

p_a Ambient pressure in hPa

ρ Tin concentration by weight per tin compound in the ambient air in µg/m^3 referring to t_a and p_a

ρ_0 Tin concentration by weight per tin compound in the ambient air in µg/m^3 at 20 °C and 1013 hPa

σ Tin concentration by volume per tin compound in the ambient air in µL/m^3

X Weight of tin per tin compound in the desorption solution in mg

8 Reliability of the method

8.1 Precision

In order to determine the precision for the complete method volumes of 200 µL each of solutions of two different concentrations of tetrabutyltin, tributyltin chloride, dibutyltin dichloride and butyltin trichloride in t-butyl methyl ether (concentration of the individual components cf. Tab. 2) were pipetted twenty times into small tubes by means of a precise dosing device (schematic representation of the apparatus cf. Fig. 2). 100 L air were drawn through each tube and another tube with the exchange resin connected behind. At the selected concentrations this corresponds with about 0.15 mg Sn/m^3 or 0.015 mg Sn/m^3 airborne organotin compounds. Each sample prepared according to Sect. 3.6 was analysed six times by gas chromatography, three times using flame ionization detection and three times using mass selective detection.

Table 2. Standard deviation (rel.) s, mean variation u and recovery rate η.

Detector	Component	Applied weight of tin	Recovered weight of tin mean of ten individual determinations	Standard deviation (rel.)	Mean deviation	Recovery rate
		X	$X*$	s	u	η
		µg Sn/ 200 µL	µg Sn	%	%	
FID	Bu$_4$Sn	2.410	2.027	2.4	5.2	0.841
	Bu$_3$SnCl	6.050	5.553	1.4	3.0	0.918
	Bu$_2$SnCl$_2$	3.370	2.870	1.4	3.2	0.852
	BuSnCl$_3$	2.995	2.402	2.1	4.7	0.802
FID	Bu$_4$Sn	0.234	0.140	6.7	14.3	0.598
	Bu$_3$SnCl	0.603	0.566	3.2	7.1	0.939
	Bu$_2$SnCl$_2$	0.342	0.305	1.4	2.9	0.892
	BuSnCl$_3$	0.299	0.229	1.4	2.9	0.766
MS	Bu$_4$Sn	2.410	1.784	5.9	13.3	0.740
	Bu$_3$SnCl	6.050	5.124	5.3	11.9	0.847
	Bu$_2$SnCl$_2$	3.370	2.145	4.0	8.9	0.717
	BuSnCl$_3$	2.995	2.376	3.2	7.1	0.793
MS	Bu$_4$Sn	0.234	0.131[1]	8.6	19.4	0.562
	Bu$_3$SnCl	0.603	0.568[1]	3.6	8.0	0.943
	Bu$_2$SnCl$_2$	0.342	0.306[1]	2.0	4.4	0.896
	BuSnCl$_3$	0.299	0.228[1]	4.1	9.1	0.761

[1] Mean of 9 individual determinations

8.2 Recovery rate

The recovery rate $\eta = X*/X$ (i.e. the ratio of the real analytically determined quantity $X*$ to the quantity X in the sample) was determined according to the experiments in Sect. 8.1. The results for the components, concentrations and detectors are given in Table 2.

8.3 Detection limit

The detection limit depends on the used apparatus. The detection limit with the measuring apparatus and the measuring parameter applied is 50 pg tin (measuring value corresponds with the threefold background noise). At a sample volume of 100 L and a final volume of the solution of 200 µL a detection limit of 0.1 µg Sn/m^3 for each individual component can be achieved.

9 Discussion

The described analytical method permits the determination of individual airborne organotin compounds in short-term measurements as well as in 8 h average measurements.
The airborne organotin compounds are separated quantitatively on glass fibre filters or on membrane filters if they appear as dust (e. g. packing of solid products) or aerosol (e. g. spraying of TBTO containing antifouling paint) [4, 6]. The adsorption of organotin compounds on activated carbon also occurs quantitatively but the desorption is very uncomplete [5, 6]. Silica gel [7] or the organic polymers Chromosorb 102 [8] or Tenax GC and Amberlite XAD-2 [6] are recommended as adsorbents. The desorption with acid containing solvents is complete.
Interferences are possible if the tetraalkyltin compounds which are only bound adsorptively to the ion exchanger are displaced from the exchange resin by organic air pollutants (e. g. solvent vapours). For measurements in working areas impacted to such an extent it is recommended to connect a second adsorption tube (XAD-2 or XAD-4). Other interferences are not known.
The application of ion exchange resins in the gas adsorption is described previously [9, 10]. Desorption, alkylation with alkylmagnesium halogenide and gas chromatographic determination even with mass spectrometric detection were published by Zimmerli [8] in accordance with the analogous procedure in the water analytics.

Apparatus:
Gas chromatograph HP 5880 equipped with flame ionization detector or
Gas chromatograph HP 5890 equipped with mass selective detector (MSD) HP 5970 B of Hewlett Packard
Autosampler HP 7673 A with on-column injection
Chromatogram registration and evaluation:
– FID: HP 5880 GC Terminal, Level 4 of Hewlett Packard
– MSD: HP 9816, Series 200, Workstation of Hewlett Packard
Pump: AMA PN 7300 of AMA, Hilden

10 References

[1] *Deutsche Forschungsgemeinschaft (DFG)* (1998) List of MAK and BAT values 1998. Maximum concentrations and biological tolerance values at the workplace. Commission for the Investigation of Health Hazards of Chemical Compounds in the Work Area, Report No 34. WILEY-VCH Verlag GmbH, Weinheim.
[2] *Seiler HG, Seiler M* (1994) Tin in urine. In: Angerer J, Schaller KH (Ed) Analyses of hazardous substances in biological materials, Vol. 4. VCH-Verlagsgesellschaft, Weinheim, 223–240.
[3] *Greim H* (Ed) (1996) Di-*n*-octyltin compounds and mono-*n*-octyltin compounds. In: Occupational Toxicants. Critical data evaluation for MAK values and classification of carcinogenes, Vol 7. VCH Verlagsgesellschaft, Weinheim, 91–114.

[4] *Jeltes R* (1969) Determination of bis(tributyltin)oxide in air by atomic absorption spectroscopy or pyrolysis gas chromatography. Ann Occup Hyg 12: 203–207.

[5] *National Institute for Occupational Safety and Health (NIOSH)* (1976) Criteria for a recommended standard: Organotin compounds. U.S. Department of Health, Education and Welfare. Pub No 77–115.

[6] *Gutknecht WF, Grohse PM, Homzak CA, Tronzo C, Ranade MB, Damle A* (1982) Development of a method for the sampling and analysis of organotin compounds. National Technical Information Service (NTIS), U.S. Department of Commerce. Pub No Pb 83-180737.

[7] *Riggle CJ, Sgontz DL, Graffeo AP* (1978) The analysis of organotins in the environment. Int. Conf. Sens. Environ. Pollution 4th (1977). Citation: CA 89: 11085.

[8] *Zimmerli B, Zimmermann H* (1980) Gaschromatographische Bestimmung von Spuren von n-Butylzinnverbindungen (Tetra-, Tri-, Di-) in Luft. Z Anal Chem 304: 23–27.

[9] *Vaidyanathan AS, Youngquist GR* (1973) Sorption of sulfur dioxide, hydrogen sulfide and nitrogen dioxide by ion exchange resins. Ind Eng Chem Prod Res Develop 12: 288.

[10] *Siemer DD* (1987) Ion exchange resins for trapping gases: Carbonate determination. Anal Chem 59: 2439.

Authors: *H.A. Müller, U. Schings-Hartmann*
Examiners: *W. Kleiböhmer, K. Cammann*

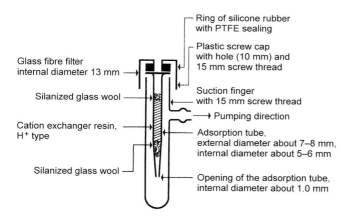

Fig. 1. Sampling apparatus.

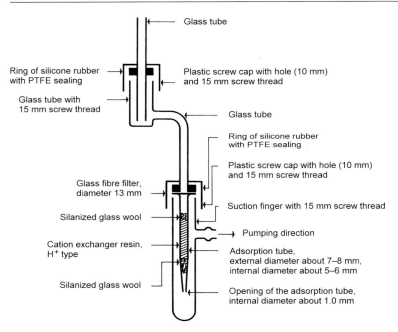

Glass tube

Ring of silicone rubber
with PTFE sealing

Glass tube with
15 mm screw thread

Plastic screw cap with hole (10 mm)
and 15 mm screw thread

Glass tube

Ring of silicone rubber
with PTFE sealing

Plastic screw cap with hole (10 mm)
and 15 mm screw thread

Glass fibre filter,
diameter 13 mm

Silanized glass wool

Suction finger with 15 mm screw thread

Pumping direction

Cation exchanger resin,
H^+ type

Adsorption tube,
external diameter about 7–8 mm,
internal diameter about 5–6 mm

Silanized glass wool

Opening of the adsorption tube,
internal diameter about 1.0 mm

Fig. 2. Apparatus for the determination of the recovery rate.

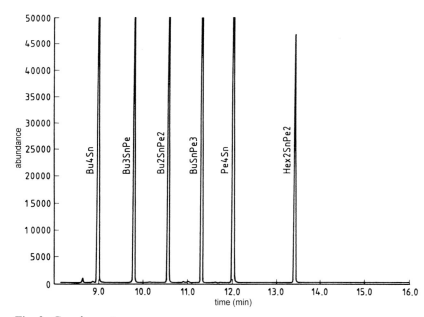

Fig. 3. Gas chromatogram.

Ozone

Method number   1

Application   Air analysis

Analytical principle   Photometry

Completed in   May 1992

Summary

The method described for the determination of ozone is based on the method proposed by the VDI for monitoring ambient air. The VDI [1, 2] method is also suitable to determine airborne ozone at the workplace. Ozone-containing air is drawn through two serially connected wash bottles equipped with frits. The decolourization of the indigo carmine solution is measured by use of photometry within 5 min.

Precision:	Standard deviation (rel.) $s = 9.7\%$
	Mean variation $u = 21.6\%$
	where $n = 11$ determinations and $c = 0.2$ mg/m^3
Sensitivity:	Reciprocal calibration factor $k' = 16.72$ µg (referring to 25 mL measuring solution and 5 cm cuvettes)
Quantification limit:	0.85 µg ozone $\triangleq$ 0.011 mg/m^3
Recommended sampling time:	up to 2 h
Recommended sample volume:	80 L

Ozone

Ozone (O_3) is an unstable, pungent smelling, extremely toxic gas (molecular weight 48.0 g/mole; melting point $-192.5\,°C$; boiling point $-111.9\,°C$). Ozone is one of the strongest oxidation agents. It is readily soluble in alcohol and oils.

In a height of 15–25 km ozone is formed of oxygen under the influence of UV rays. At ground levels concentrations of 0.02–0.05 ppm are measured. In addition it is formed during industrial processes which generate great intensities of UV rays (e. g. arc welding).

Ozone is used in the bleaching of oils, synthetic fibres, paper, cellulose and textiles. The most important range of application is the processing of drinking water according to DIN 19627.

Ozone is listed in the carcinogenic category III 3 [3]. The currently valid TRK value (1998) is 0.2 mg/m^3 and 0.1 mL/m^3 [4].

Author: *D. Breuer*
Examiner: *Th. zur Mühlen*

Ozone

Method number   1

Application   Air analysis

Analytical principle Photometry

Completed in   May 1992

Contents

1 General principles

The method described for the determination of ozone is based on the method proposed by the VDI for monitoring ambient air. The VDI [1, 2] method is also suitable to determine airborne ozone at the workplace. Ozone-containing air is drawn through two serially connected wash bottles equipped with frits. The decolourization of the indigo carmine solution is measured by use of photometry within 5 min.

2 Equipment, chemicals and solutions

2.1 Equipment

Pump, pumping capacity 40 L/h
Ozone generator
Gasmeter
Photometer suitable to measure in the wavelength of about 623 nm
Wash bottles with frits
2 or 5 cm Cuvettes
25 mL Bulb pipette
500, 1000 mL Volumetric flasks
100 mL Graduated cylinder
Polytetrafluoroethylene (PTFE) hoses
Stopwatch
Thermometer
Barometer

2.2 Chemicals

Indigo carmine, analytical grade, disodium salt of the 5,5′-indigosulfonic acid
Disodium phosphate, analytical grade, $Na_2HPO_4 \cdot 12\ H_2O$
Monopotassium phosphate, analytical grade, KH_2PO_4

2.3 Solutions

Solution I:
In a 1000 mL volumetric flask 28 g $Na_2HPO_4 \cdot 12\ H_2O$ and 80 g KH_2PO_4 are dissolved in distilled water and diluted to the mark with distilled water. The pH-value of the solution is adjusted to 5.5.

Solution II:
In a 500 mL volumetric flask 150 mg indigo carmine are completely dissolved, 50 mL of solution I are added and the solution is diluted up to the mark with distilled water.

Solution III:
20 mL of solution II and 100 mL of solution I are mixed in a 1000 mL volumetric flask and diluted up to the mark with distilled water. For the analysis the solution has to be freshly prepared. The shelf life of the solution is 24 h.

3 Calibration

The method is calibrated by parallel measurements according to the basic method [2]. The generation of the ozone containing air occurs by means of an ozone generator. The measurements are carried out as described in Sect. 4. The ozone concentration is adjusted at the ozone generator; different calibration concentrations can be adjusted by varying the sampling times. The decolourization of the indigo carmine solution of **both** wash bottles is measured.

From the sum of the decolourization $\Sigma \Delta E = \Delta E_1 + \Delta E_2$ with $\Delta E_1 = E_{0,1} - E_1$ and $\Delta E_2 = E_{0,2} - E_2$ the decrease of the extinction is plotted against the absolute ozone quantity to obtain the calibration curve.

The calibration curve was made for the range of 1–12 µg ozone. The calibration factor – referring to a 5 cm cuvette – is $k = 0.0598$ µg , the reciprocal calibration factor is k' = 16.72 µg^{-1}. The correlation coefficient of the calibration curve is $r = 0.995$. An example of a calibration curve is shown in Fig. 1.

4 Sample collection and analytical determination

25 mL each of solution III are transferred into two serially connected wash bottles with frits (cf. Sect. 2.3) and installed in the sampling device (cf. Fig. 2). By a flow-stabilized pump or a pump controlled by a gasmeter a flow rate of 40 L/h is applied (sampling time up to 2 h). If higher concentrations are observed (rapid decolourization of the solution) the sampling time has to be reduced.

After the sampling the wash bottles are closed airtight and analysed within 5 min. In the laboratory the solutions are transferred separately into 5 cm cuvettes and the decolourization is determined.

Under the conditions as described the absorption in the first wash bottle is greater than 90%.

5 Calculation of the analytical result

For the calculation of the airborne ozone concentration at the workplace the following equations are applied:

$$\rho = \frac{\Sigma \Delta E \cdot k'}{V_Z} \cdot \frac{273 + t_g}{273 + t_a}$$

At 20 °C and 1013 hPa:

$$\rho_0 = \rho \, \frac{273 + t_a}{293} \cdot \frac{1013 \, \text{hPa}}{p_a}$$

The corresponding concentration by volume σ – independent of pressure and temperature – is:

$$\sigma = \rho_0 \, \frac{24.06 \, \text{L/mole}}{48.00 \, \text{g/mole}}$$

$$\sigma = \rho \cdot \frac{273 + t_a}{p_a} \cdot \frac{1013 \, \text{hPa}}{293} \cdot \frac{24.06 \, \text{L/mole}}{48.00 \, \text{g/mole}}$$

At $t_a = 20\,°\text{C}$ and $p_a = 1013 \, \text{hPa}$:

$$\sigma = \rho \cdot \frac{273 + t_a}{p_a} \, 1.733 \, \frac{\text{hPa} \cdot \text{mL}}{\text{mg}}$$

$$\sigma = \rho \cdot 0.501 \, \frac{\text{mL}}{\text{mg}}$$

Legend:

$\Sigma \Delta E = \Delta E_1 + \Delta E_2$	Sum of the decrease of the extinction of the indigo carmin solutions
k'	Reciprocal calibration factor in mg
V_Z	Sample volume in m^3
t_a	Temperature of the ambient air in $°\text{C}$
t_g	Temperature in the gasmeter in $°\text{C}$
p_a	Pressure of the ambient air in hPa
ρ	Airborne ozone concentration by weight in mg/m^3 referring to t_a and p_a
ρ_0	Airborne ozone concentration by weight in mg/m^3 referring to $20\,°\text{C}$ and 1013 hPa
σ	Airborne ozone concentration by volume in mL/m^3

6 Reliability of the method

6.1 Precision

To test the reproducibility of the method defined ozone concentrations were adjusted at the ozone generator and drawn through the absorption solution. The concentrations $0.1 \, \text{mg/m}^3$, $0.2 \, \text{mg/m}^3$ and $0.4 \, \text{mg/m}^3$ were applied (cf. Tab. 1).

Table 1. Standard deviation (rel.) s.

Concentration mg/m^3	Number of individual measurements n	Standard deviation (rel.) s %
0.1	12	16.5
0.2	11	9.7
0.4	9	11.2

6.2 Determination of the blank value and the quantification limit

The quantification limit for ozone was calculated from the standard deviation of the blank value $M_b = E_0$ as follows:

$$C = M_b + 3s_0$$

C Relative quantification limit
s_0 Standard deviation of the blank value
M_b Average blank value

The values $M_b = 1.1286$ and $s_0 = 0.0067$ were determined from a total number of 16 individual measurements. This leads to an absolute quantification limit of 0.85 µg ozone. The relative quantification limit is 0.011 mg/m^3 for a 80 L sample volume.

6.3 Specificity

The method has no cross sensitivity to SO_2. The presence of NO_2 effects slightly higher concentration values.

6.4 Shelf life

The time between sampling and analysis should not exceed 5 min.

7 References

[1] *Verein Deutscher Ingenieure (VDI)* (1978) VDI-Richtlinie 2468 Blatt 5, Messen der Ozon-konzentration – Indigosulfonsäure-Verfahren. Beuth Verlag, Berlin.
[2] *Verein Deutscher Ingenieure (VDI)* (1978) VDI-Richtlinie 2468 Blatt 1, Messen der Ozon-konzentration – Kaliumjodid-Methode (Basisverfahren). Beuth Verlag, Berlin.
[3] *Deutsche Forschungsgemeinschaft* (1998) List of MAK and BAT values 1998. Maximum concentrations and biological tolerance values at the workplace. Report No 34 by the Com-

mission for the Investigation of Health Hazards of Chemical Compounds in the work area.
WILEY-VCH Verlag GmbH, Weinheim.

[4] *Bundesministerium für Arbeit und Sozialordnung* (1998) TRGS 900: Grenzwerte in der Luft am
Arbeitsplatz-Luftgrenzwerte. Technische Regeln und Richtlinien des BMA zur Verordnung über
gefährliche Stoffe. BArbBl. 10/1996, p 106–128, last supplement 10/1998, p 73–74.

Author: *D. Breuer*
Examiner: *Th. zur Mühlen*

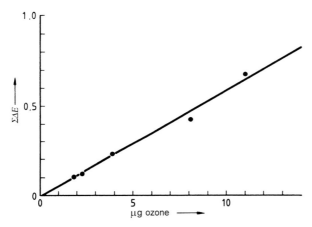

Fig. 1. Calibration curve ($\Sigma \Delta E = \Delta E_1 + \Delta E_2$).

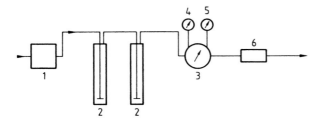

Fig. 2. Calibration apparatus.

1 Ozone generator
2 Wash bottles with frits
3 Gasmeter
4 Manometer
5 Thermometer
6 Pump

Phenol

Method number   2

Application   Air analysis

Analytical principle Gas chromatography

Completed in   November 1994

Summary

The method permits the determination of airborne phenol concentrations at workplaces as a shift average value and as a peak value.
The phenol vapour is adsorbed on silica gel in adsorption tubes (Dräger, type B), desorbed with diethyl ether, silylated and analysed in this solution by gas chromatography. The evaluation makes use of undecane as internal standard.

Precision of the gas chromatographic analysis:	Standard deviation (rel.) $s = 0.95\%$ Mean variation $\quad u = 2.4\%$ at a phenol concentration of 2.1 mg/m^3 and a sampled air volume of 40 L and $n = 6$ duplicate determinations
Precision of the complete method:	Standard deviation (rel.) $s = 3.7$ and 2.0% Mean variation $\quad u = 9.5$ and 5.1% at phenol concentrations of 1 and 40 mg/m^3 and a sampled air volume of 40 L and $n = 6$ duplicate determinations
Recovery rate:	$\eta = 0.87$ (87%)
Detection limit:	4 ng absolute 0.5 mg/m^3 for a sampled air volume of 40 L and 5 mL of desorption agent
Recommended sampling time:	2 h
Recommended sample volume:	40 L

Phenol

C₆H₅–OH

Phenol (molecular weight 94.11 g/mole, melting point 40 °C) forms colourless prisms and has a characteristic smell. Phenol is highly corrosive on the skin and mucous membranes. Inhaling phenol vapour may result in respiratory tract damage.

The intake of high phenol concentrations leads to paralysis of the central nervous system. Phenol ist listed in the carcinogenic category III 3 [1]. The currently valid TRK value (1998) is 19 mg/m³ and 5 mL/m³ [2].

Phenol is used mainly in the production of phenol formaldehyde resins.

Authors: *M. Hennig, N. Lichtenstein*
Examiner: *U. Knecht*

Phenol

Method number   2

Application   Air analysis

Analytical principle   Gas chromatography

Completed in   November 1994

Contents

1 General principles

To determine the concentration of phenol in the workplace air measured volumes of air are drawn through silica gel in adsorption tubes (Dräger, type B). The phenol adsorbed on the silica gel is desorbed with diethyl ether, silylated and analysed by means of gas chromatography. The quantitative evaluation is carried out using an internal standard.

2 Equipment, chemicals and solutions

2.1 Equipment

Gas chromatograph equipped with flame ionization detector (FID)
Capillary column
Integrator/computer
Flow stabilized sampling pump (flow rate 20 L/h)
Silica gel adsorption tubes (e. g. Dräger, type B)
Gasmeter
Microliter syringes
5 mL Pipettes
7 mL Vials or screw-cap glass vessels
Vials for an autosampler and microadapters
5 mL Volumetric flasks

2.2 Chemicals

Phenol, analytical grade
n-Undecane, analytical grade, internal standard
N,O-Bis[trimethylsilyl]trifluoroacetamide (BSTFA)
Diethyl ether, analytical grade

2.3 Solutions

Phenol stock solution:
About 50 mg of phenol are weighed exactly into a 5 mL volumetric flask. Then the flask is filled up to the mark with diethyl ether (10 g/L).

2.4 Calibration standards

0.5 µL *n*-undecane (IS) is pipetted into each of five 5 mL volumetric flasks containing about 2 mL of diethyl ether. The calibration standards are prepared according to the pipetting scheme in Table 1. For this purpose 2–240 µL of the stock solution is added to the *n*-undecane and the flasks are filled up to the mark with diethyl ether. (Phenol weights and concentrations are given for exactly 50 mg of the weighed phenol sample).

Table 1. Pipetting scheme for the preparation of calibration standards.

Volume of the stock solution µL	Final volume of the calibration standards mL	Phenol concentration in the calibration standard mg/L	Phenol concentration in air for 40 L air samples mg/m^3
2	5	4	0.5
8	5	16	2
20	5	40	5
80	5	160	20
240	5	480	60

Using these calibration standards and sampling an air volume of 40 L, phenol concentrations in the range between 0.5–60 mg/m^3 can be determined.
The calibration standards are derivatized as described in Sect. 3.2.

3 Sample collection and preparation

3.1 Sample collection

For two hours and at a flow rate of 20 L/h air is drawn through silica gel in an adsorption tube which is connected to a sampling pump. When sampling is complete the flow rate is measured again. The deviation from the value set at the start of sampling should not be greater than 5%. Then the tubes are closed and the experimental sampling parameters (temperature, sampling time, flow rate) are noted.

3.2 Sample preparation

The contents of a silica gel adsorption tube are transferred into a sealable 7 mL vial and 5 mL of diethyl ether is added. Generally it is not necessary to separate the collection phase from the control phase.
If the two phases are separated, the short phase is eluted with 2 mL and the long phase with 3 mL diethyl ether. 0.1 µL *n*-undecane per mL is added as internal standard. The closed vials are allowed to stand until the next day. 100 µL of the sample is silylated

with 5 μL of BSTFA by warming with the reagent to about 60 °C for 30 min. If an autosampler is used, microadapters are recommended.

4 Operating conditions for gas chromatography

Column:	Material:	Fused silica
	Length:	30 m
	Internal diameter:	0.25 mm
Stationary phase:	DB 5 from J & W	
	(5% phenylsilicone and 95% methylsilicone)	
	Film thickness:	0.25 μm
Detector:	Flame ionization detector	
Temperatures:	Oven:	50 °C for 10 min isothermal, then 5 °C/min up to 150 °C
	Injector:	200 °C
	Detector:	220 °C
Carrier gas:	Helium	
	Column head pressure: 100 kPa	
	Split:	1 : 15
Detector gases:	Hydrogen	
	Synthetic air	
Injection volume:	1 μL	
Analysis time:	30 min	

An example of a gas chromatogramm is shown in Fig. 1.

5 Analytical determination

A volume of 1 μL of the derivatized sample solution is injected at least twice into the gas chromatograph. The evaluation is carried out using a calibration curve in which the ratios of the peak areas for the phenol derivatives and the internal standard are plotted against the weight of phenol (cf. Fig. 2). When a computerized evaluation is used, the weight of phenol is determined from a calibration function calculated by the computer.

6 Calibration

The calibration standards are silylated as described in Sect. 3.2 and each solution is analysed three times under the conditions described above. The ratios of the peak areas for the phenol derivatives and the internal standard are plotted against the weights of phenol used. The calibration function is linear in the indicated concentration range. If a computer is used for the evaluation the calibration curve, that is the calibration function, is calculated from the peak areas and the corresponding weights by the computer.

7 Calculation of the analytical result

The concentration value (mg/m^3) is obtained directly from the computerized evaluation if the sampled air volume and the phenol weight are included in the calculation. If the evaluation is carried out using a calibration curve, the phenol weights corresponding to the particular ratios are read from the calibration curve.
The concentration by weight is calculated with the following equation:

$$\rho = \frac{X}{V \cdot \eta}$$

Legend:
ρ Concentration of phenol in the air (w/v)
X Weight of phenol in the elution solution
V Sampled air volume in L at 20 °C and 1013 hPa
η Recovery rate

8 Reliability of the method

8.1 Precision

To check the reliability of the gas chromatographic determination six calibration solutions were prepared from the stock solution and each analysed twice. At a concentration of 2.1 mg/m^3 and a sampled air volume of 40 L a standard deviation (rel.) of 0.95% was obtained. The mean variation was 2.4%.
To determine the precision of the complete method six silica gel adsorption tubes were loaded with a measured weight of phenol, prepared and analysed according to the above instructions. The low concentration (1 mg/m^3 for a sampled air volume of 40 L) was injected directly into the tubes as a diethyl ether solution. At the higher phenol concentration (40 mg/m^3 and a sampled air volume of 40 L) the tubes were loaded from the gas phase. For this purpose an appropriate volume of phenol solution was placed in a gas collection vessel equipped with a septum and connected to a silica gel adsorption tube and a pump. At a flow rate of 20 L/h, the tube was loaded for two hours. The gas collection vessel was warmed up after 1 h (hot air) to ensure that the phenol evaporated completely.

Duplicate analysis of the phenol in the tubes yielded a standard deviation of 3.7% and a mean variation of 9.5% at a phenol concentration of 1 mg/m^3 (for a sampled air volume of 40 L) and a standard deviation of 2.0% and a mean variation of 5.1% at a phenol concentration of 40 mg/m^3 (for a sampled air volume of 40 L).

8.2 Recovery rate

The desorption efficiency is independent of concentration in the given concentration range. It amounts to 87%. The desorption efficiency was calculated as a ratio of the phenol weights detected from loaded tubes and the calibration standards containing the same weights of phenol.

8.3 Shelf life

The loaded silica gel adsorption tubes can be stored at room temperature for at least 14 days without loss of phenol.

8.4 Detection limit

Under the described experimental conditions the detection limit is 0.5 mg/m^3.

8.5 Specificity

Interferences by compounds with the same retention time as the phenol derivative are possible. Generally the separation of interfering components is not difficult because of the relatively long retention time of the phenol derivative.

9 Discussion

The method described permits precise determination of airborne phenol in the range of about 1/40 up to four times the currently valid TRK value (1998).

As a rule, a break-through of the phenol through the collection phase of the adsorption tube is unlikely up to concentrations equivalent to four times the currently valid TRK value. Therefore, separation of the collection phase and the control phase is generally unnecessary. Besides phenol, cresol compounds may also be determined by this method. The analysis time can be reduced to 22 min if phenol is the only analyte. Variation of the temperature program may result in shorter analysis times.

Silylation reagents other than BSTFA may be used. The derivatization of phenol by acetic anhydride/pyridine has also been found to be satisfactory by the examiner who was also able to replace the highly volatile diethyl ether by acetonitrile.

Apparatus used:
Gas chromatograph SICHROMAT from Siemens

10 References

[1] *Deutsche Forschungsgemeinschaft* (1998) List of MAK and BAT values 1998. Maximum concentrations and biological tolerance values at the workplace. Report No 34 by the Commission for the Investigation of Health Hazards of Chemical Compounds in the work area. WILEY-VCH Verlag GmbH, Weinheim.
[2] *Bundesministerium für Arbeit und Sozialordnung* (1998) TRGS 900: Grenzwerte in der Luft am Arbeitsplatz – Luftgrenzwerte. Technische Regeln und Richtlinien des BMA zur Verordnung über gefährliche Stoffe. BArbBl. 10/1996, p 106–128, last supplement 10/1998, p 73–74.

Authors: *M. Hennig, N. Lichtenstein*
Examiner: *U. Knecht*

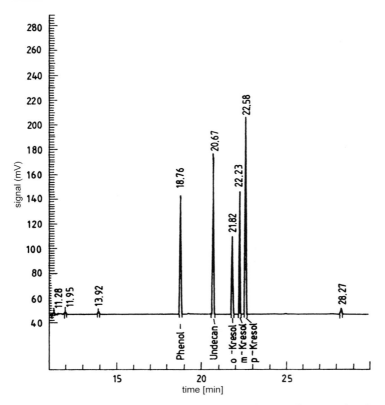

Fig. 1. Gas chromatogram of airborne phenol and *o*-cresol, *m*-cresol and *p*-cresol.

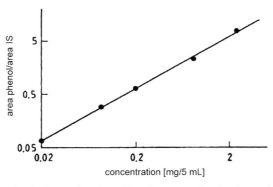

Fig. 2. Example of a calibration curve for the determination of airborne phenol.

Polyisocyanates, Aliphatic Isocyanate Prepolymer
(based on HDI)

Method number	1
Application	Air analysis
Analytical principle	High performance liquid chromatography
Completed in	June 1992

Summary

By use of a sampling pump the airborne aerosols containing the hexamethylene diisocyanate (HDI) prepolymer (polyisocyanates) are drawn through a sorption tube filled with glass wool which is impregnated with a secondary amine as a derivatization agent. The aerosol droplets are collected on the glass wool. The inlet velocity of the sampled air has to be adjusted in order to collect particles conforming to DIN EN 481 [1]. Part of the prepolymer reacts with the reagent on the filter to form soluble urea derivatives. Immediately after the sampling the filter is back-eluted with dichloromethane. Any unreacted free isocyanate groups rapidly react in the homogeneous phase to urea derivatives; the total being analysed by high performance liquid chromatography with UV detection. The quantitative evaluation occurs by a calibration curve. Calibration is carried out with a sample of the bulk prepolymer as present in the workplace sample. The prepolymer is a constituent in the so-called hardener of two-component systems with about 40% by weight resp. with > 70% in one-component systems. The conversion with the derivatization agent occurs according to the Meth. HDI, TDI, Vol. 1, p. 67 [2].

Measuring range:	$0.1-10$ mg/m^3 airborne polyisocyanate
Precision:	Standard deviation (rel.) $s = 6.8\%$
	Mean variation $u = 17.3\%$
	at an applied quantity of 2.4 mg and $n = 5$ determinations
Quantification limit:	100 µg Polyisocyanate per m^3 air (corresponding with 3 µg per mL sample solution at a sample volume of 30 µL, 1 mL sample solution and 25 µL injection volume)

Recovery rate: $\eta = 1.06$ (106%)
Recommended sampling time: 15 min
Recommended sampling volume: 30 L

Polyisocyanate

Polyisocyanates formally are incomplete conversion products of monomeric diisocyanates with substances containing low-molecular bifunctional or trifunctional hydroxyl groups (or amine groups). They form prepolymeric polyfunctional isocyanates with urethane structures, biuret structures or isocyanate (trimerisate) structures [3]. In order to obtain three dimensional cross-linked high-quality surface films (lacquers) the idealized structures are at least trifunctional prepolymers of the corresponding monomeric diisocyanates. In addition, besides traces of the initial isocyanate from secondary reactions they contain small amounts of dimers but mainly higher-molecular weight components. Therefore the chromatogram of the derivatized prepolymer is a heterogeneous but characteristic peak pattern in which the low-molecular main constituents with an ideal structure dominate (cf. Fig. 1 and structure examples). The peak pattern is possibly influenced by solvent components of a similar retention behaviour. For this reason in the analytics the peak pattern has to be considered for the qualitative and quantitative determination. The industrially most important isocyanates for spraying applications are products based on the aliphatic hexamethylene diisocyanate (HDI).
Products on the basis of the aromatic toluene diisocyanate (TDI, an isomeric mixture of 2,4-TDI and 2,6-TDI) or mixed systems of both monomeric types are also used.
The described method deals only with polyisocyanates based on the aliphatic hexamethylene diisocyanate.
The trademarks of the polyisocyanates are:

Desmodur® (Bayer)
Basonat® (BASF)
Tolonate® (Rhone-Poulenc, France)
Mondur® (Mobay, USA)

They differ in their type and they contain different quantities of solvents. The solvent quantity is only an application related problem and is generally assumed as not known. Due to low demand and limited storage life the pure polyisocyanates are only obtainable in large industrial quantities, but not from laboratory chemical suppliers.

Polyisocyanate based on HDI with biuret structure (ideal structure)

Desmodur N (Bayer), Type 3200

Polyisocyanate based on HDI with isocyanate structure (ideal structure of the trimerisate)

$$O=C=N-(CH_2)_6-N \underset{O}{\overset{O}{\bigotimes}} \begin{matrix} (CH_2)_6-N=C=O \\ N \\ =O \\ N \\ (CH_2)_6-N=C=O \end{matrix}$$

Desmodur N (Bayer), Type 3300

Explanations

The spraying of polyisocyanate containing reaction mixtures causes aerosol droplets in the area of exposure. Besides the polyisocyanates the mixture generally contains a small quantity of monomeric isocyanate (below 0.5%, relative to the polymeric isocyanate content). Compared with the monomeric part in the droplet-shaped isocyanate the quantity of the monomeric vaporous diisocyanates can be neglected. The described method permits the determination of the airborne concentration of the polyisocyanates at workplaces. It can also be applied to determine the monomeric diisocyanates in airborne droplet-aerosols at workplaces (cf. Meth. HDI, TDI, Vol. 1, p. 67 [2]).
The method is not appropriate for very rapid reaction systems in which a very rapid reaction of the isocyanate groups is expected during the sampling time, i. e. they can react within the range of minutes to form unsoluble polymer structures. For reasons of handling the potlifes of the conventional lacquer systems on the above mentioned basis are generally within the range of hours whereas the application related sampling times are in the range of minutes. Even after exceeding the potlife of more than 100% these lacquer systems are still soluble. The reaction degree of the isocyanates is below 10%, i.e within the error range of the complete method. In order to test the applicability of the method a sample can be taken from the reaction batch before and after the processing directly followed by derivatization. In the subsequent analysis the polyisocyanate concentration in both samples must be the same within the tolerance range.

Author: *M. Kuck*
Examiner: *J. U. Hahn*

Polyisocyanates, Aliphatic Isocyanate Prepolymer
(based on HDI)

Method number   1

Application   Air analysis

Analytical principle High performance liquid chromatography

Completed in   June 1992

Contents

1 General principles

By use of a sampling pump the airborne aerosols containing the hexamethylene diiso-cyanate (HDI) prepolymer (polyisocyanates) are drawn through a sorption tube filled with glass wool which is impregnated with a secondary amine as a derivatization agent. The aerosol droplets are collected on the glass wool. The inlet velocity of the sampled air has to be adjusted in order to collect particles conforming to DIN EN 481 [1]. Part of the prepolymer reacts with the reagent on the filter to form soluble urea derivatives. Immediately after the sampling the filter is back-eluted with dichloro-methane. Any unreacted free isocyanate groups rapidly react in the homogeneous phase to urea derivatives; the total being analysed by high performance liquid chroma-tography with UV detection. The quantitative evaluation occurs by a calibration curve. Calibration is carried out with a sample of the bulk prepolymer as present in the work-place sample. The prepolymer is a constituent in the so-called hardener of two-compo-nent systems with about 40% by weight resp. with >70% in one-component systems. The conversion with the derivatization agent occurs according to the Meth. HDI, TDI, Vol. 1, p. 67 [2].

Depending on the calibration and on the evaluation the analysis result can be given as polyisocyanate or total isocyanate in the air at workplaces.

2 Equipment, chemicals and solution

2.1 Equipment

High pressure liquid chromatograph equipped with UV diode array detector
Separation column: ET 250/8/4 Nucleosil 120–5 C_8
10 mL Test-tubes with polyethylene stoppers
Various volumetric flasks
500 mL Separation funnel
Various measuring pipettes
Device for concentrating under nitrogen
Device for concentrating under vacuum

Flow stabilized membrane pump, minimum flow rate 2 L/min (e.g. GSA 2000 from Gesellschaft für Staubmeßtechnik und Arbeitsschutz mbH, Neuss)

Calibration device for the calibration of the flow rate of the pump, suitable for flow rates of 2 L/min

Tablet tubes, preparation vials (e.g. 40 mL EPA vials from ASS-CHEM, Bad Homburg)

Thermometer

Barometer

Tube material of glass (8 mm external diameter) for the preparation of the adsorption tubes, length 70–80 mm corresponding with an internal diameter of about 6 mm, filled with about 0.5 g long-fibre purified glass wool for an adsorption zone of 50 mm

2.2 Chemicals

Methanol, analytical grade
Toluene, analytical grade
Dichloromethane, analytical grade
Sodium sulfate, analytical grade
N-4-Nitrobenzyl-N-n-propylamine hydrochloride (nitro-reagent) (e.g. from Riedel de Haen or DFG Meth. HDI, TDI [2])
1 N Sodium hydroxide, analytical grade
Triethylamine, analytical grade
Orthophosphoric acid, analytical grade, 85%
Purified glass wool

2.3 Solutions

2.3.1 Buffer solution for HPLC

40 mL triethylamine and 20 mL orthophosphoric acid are transferred into a 1 L volumetric flask and it is diluted up to the mark with distilled water.

2.3.2 Nitro-reagent stock solution

0.8 g nitro-reagent hydrochloride (or 0.67 g of the free base) are dissolved in 50 mL distilled water and 25 mL 1 N NaOH are added. A white precipitate is formed (free base). The aqueous suspension is transferred into a 500 mL separation funnel and extracted with 200 mL dichloromethane. The organic phase is separated, dried over sodium sulfate and transferred into a 250 mL measuring flask which is then filled up to the mark with dichloromethane. The flask is encased with an aluminium foil as a light protection. This stock solution contains 2.7 mg nitro-reagent/mL. It has to be stored in a refrigerator and used within one week.

2.4 Calibration standard

2.4.1 Calibration standard for the qualitative analysis

5 mL of the collected sample material (cf. Sect. 3.1) is treated in a vacuum evaporator at a maximum temperature of 50 °C in the presence of inert gas under exclusion of moisture in order to evaporize the volatiles. One drop of the remaining material is converted with the nitro-reagent stock solution (cf. Sect. 2.4.2).

2.4.2 Calibration standard for the quantitative analysis

In one-component systems one drop of the material and in two-component systems five drops of the curing agent, correspond to about 50 mg polyisocyanate (cf. Sect. 3.1), are pipetted into a 50 mL volumetric flask and weighed. The flask is filled up to the mark with nitro-reagent stock solution (135 mg nitro-reagent/50 mL). Calibration standards containing about 50–500 µg isocyanate are prepared by diluting this solution with dichloromethane. If a curing agent of unknown origin is used as a calibration standard its polyisocyanate content must be determined separately e. g. by determination of the evaporation residue or by titration of the free NCO groups with dibutylamine according to DIN EN ISO 11 909 [4].

2.5 Sorption tubes

2.5.1 Preparation of the tubes

8 cm pieces were cut from commercial glass tubing with an external diameter of 8 mm. One end of each piece reduced by melting with a gas torch to an opening diameter of 6 mm. The tubes are filled with 500 mg long-fibre glass wool. This filling is compressed to a total length of 5 cm to form a tight cylindrical layer close to the glass wall. The distance between the narrowed opening and the cylindrical fibre layer should be only about 5 mm.

2.5.2 Impregnation of the tubes

A volume 1 mL of the nitro-reagent solution is pipetted into each tube. The tube is rotated so that the glass fibre layer and the wall are completely wetted. Then the solvent is evaporated from the tube by slightly evacuating at ambient temperature and the system flushed with inert gas. The tubes are transferred into tablet tubes which are filled with inert gas, closed gas-tight and stored under light protection. They are stable for at least 4 weeks.

3 Sample collection and preparation

3.1 Reference standards

The curing agent used in the working process (two-component system) or the readily prepared reaction system (one-component system) are filled into a glass vessel (e. g. tablet tube) which can be sealed gas-tight. The vessel is completely filled and closed with a polyethylene stopper. This sample serves as a reference material for the subsequent determination of the polyisocyanates or the isocyanate groups as well as for the preparation of the calibration standards (cf. Sect. 2.4).

3.2 Control sample (before the sample collection)

This sample is only required if the polyisocyanate reaction system is set to a potlife of less than one hour.
Immediately before the air sampling 10 mL reagent solution are transferred into a tablet tube and 1 drop of the prepared reaction batch is added and homogenized by shaking. The tube is closed gas-tight with a polyethylene cap.

3.3 Sample collection

The sampling occurs by use of adsorption tubes containing a layer of long-fibre glass wool impregnated with nitro-agent. The aerosol constitutents are adsorbed on the layer and partially chemisorbed. Immediately after the sampling the isocyanate with parts of the aerosol are back-eluted into a gastight tube by a solvent. This permit a complete reaction between the isocyanates and the reagent in a homogeneous phase and the transportation of the reaction solution into the laboratory.
For the sample collection the wide ends of the tubes are connected to the sampling pump. The sampling flow rate is adjusted to 2 L/min. At this pumping rate the internal diameter of the tube of 6 mm causes a minimum linear suction velocity of 1.25 m/s for the aerosol droplets and thus complies with the general conditions for sampling the inspiration rate according to DIN EN 481 [1].
The sampling period must be 15–30 minutes. Directly after the sampling the loaded tube is replaced with its suction side ahead into the tablet tube. On the one hand this procedure is carried out to obtain the desired quantification limit and on the other hand to exclude a falsified reaction of the isocyanates to be determined. 5 mL methylene chloride are pipetted into the free opening of the tube using a disposable pipette. The tube is closed gas-tight. The reagent elutes the glass fibre layer in a back-flush mode and causes a complete conversion of the polyisocyanates in a homogeneous phase.

3.4 Control sample (after the sample collection)

Immediately after the air sampling 10 mL of the reagent solution and 1 drop of the remaining reaction batch used for the working procedure are transferred into another tablet tube and homogenized by shaking. The tube is closed gas-tight with a polyethylene cap.

4 Operating conditions for high performance liquid chromatography

Apparatus:	HP 1090 Liquid chromatograph
Separation column:	Macherey and Nagel ET 250/8/4 NUCLEOSIL 120–5 C_8
Mobile phase:	Eluent A: Buffer solution
	Eluent B: Methanol
Gradient:	From 70% of B to 100% of B within 15 min
Flow rate:	1.8 mL/min
Detector:	UV detector, wave length: 260–272 nm
Injection volume:	25 µL

5 Analytical determination

In the laboratory the sample solution (preparation cf. Sect. 3.3) is quantitatively transferred into a glass vessel with methylene chloride. The vessel must be suitable for concentrating the solution under vacuum or by nitrogen to exactly 1 mL. After the concentration stage the operating conditions are set on the liquid chromatograph. In order to obtain the calibration curve for each measuring point 25 µL of the corresponding diluted calibration solutions are subsequently injected three times (cf. Sect. 2.4.2). The mean value of the standardized peak areas is used for each calibration point. The calibration standards are injected regularly with each batch of samples. At an assumed control value e. g. 1 mg polyisocyanate per m^3 of air the corresponding concentration of the calibration solution is 30 mg polyisocyanate per liter.

6 Calibration

6.1 Determination of the peak profile of the polyisocyanates

The calibration solutions prepared according to Sect. 2.4.1 and 2.4.2 are directly analysed and the peak patterns are compared visually. The qualitative standard contains the peaks for the assigned polyisocyanates. The corresponding peaks of the quantitative standards can be used for the quantitative analysis if they are eluted free of interference. It depends on the polyisocyanate concentration in the solution which peak is used for the evaluation. If the concentration is much higher than the quantification limit all peaks free of interference can be used for evaluation but the relative intensities may not differ by more than one order of magnitude. The determination can be limited to the most intense peak if the concentration is near the quantification limit. The certainty of the result is proportional to the number of peaks in the characteristic peak pattern with respect to the identification of the polyisocyanates. The identity of a polyisocyanate can not be confirmed on the basis of a single considered peak.

6.2 Control samples

The chromatograms of the control samples (before and after sample collection) are analysed qualitatively. It is tested by visual comparison whether the peak pattern is significantly different. If a minimum of three intense peaks occur in the chromatogram a mean value and the corresponding standard deviation can be calculated from the control samples by forming the peak area relationships. A significant change of the reaction batch during the working process can be assumed if this value is above 20 %.

6.3 Air samples

The analysis result is calculated from the peak areas which were obtained as described in Sect. 5 according to the method of the external standard by use of a calibration curve.

7 Calculation of the analytical result

According to the method of the external standard and single-point calibration under consideration of the polyisocyanate concentration of the calibration standards the analytical result is calculated from the obtained peak areas (cf. Sect. 5) as follows:

$$X = \frac{SF_{Air} \cdot G_{Ref}}{SF_{Ref}}$$

Legend:

SF_{Air} Sum of the peak areas of the most intense peaks of the air sample which are clearly to be assigned to the polyisocyanate

S_{Ref} Sum of the peak areas of the calibration standard of the components which were used for the SF_{Air}

G_{Ref} Weight of polyisocyanate corresponding with S_{Ref} in μg/L calibration solution

X Weight of polyisocyanate in the sample in μg

The corrected polycyanate concentrations by weight are calculated by the following equations:

$$\rho = \frac{X}{V_Z \cdot \eta} \cdot \frac{273 + t_g}{273 + t_a}$$

At 20 °C and 1013 hPa:

$$\rho_0 = \rho \, \frac{273 + t_a}{293} \cdot \frac{1013 \, \text{hPa}}{p_a}$$

Legend:

ρ Airborne polyisocyanate concentration by weight in mg/m³ at t_a and p_a
V_z Applied sample volume in L
η Recovery rate
ρ_0 Airborne polyisocyanate concentration by weight in mg/m³ at 20 °C and 1013 hPa
t_g Temperature in the gasmeter in °C
t_a Temperature in the sample air in °C
p_a Ambient pressure during the sampling in hPa

8 Reliability of the method

8.1 Precision and recovery rate

As described before (cf. Sect. 3) a defined amount of polyisocyanate corresponding with 2.38 mg was given on the front zone of each of five sorption tubes. Then 30 L laboratory air were drawn through each tube with a flow rate of 2 L/min. The tubes were prepared and analysed as described (for the values cf. Table 1). The standard deviation s was 0.17 mg (6.8%) and the mean variation was $u = 0.48$ mg (17.3%). The average recovery rate was 1.06 (106%).

Table 1. Recovery rate.

Sample	Theoretical amount mg	Recovered amount mg	Recovery rate %
1	2.38	2.31	97.5
2	2.38	2.66	118
3	2.38	2.66	118
4	2.38	2.63	110
5	2.38	2.57	109
Mean value	2.38	2.53	106

8.2 Quantification limit

Under the mentioned analysis conditions the quantification limit is 0.1 mg polyisocyanate per m^3 air at a sample volume of 30 L.

9 Discussion

The application range is not appropriate for rapid reaction systems in which a considerable reaction of the isocyanate groups to unsoluble polymer structures is expected during the sampling. In general, the potlifes of these lacquer systems mentioned above are sufficiently long (about 1 hour). In contrast, the practice related sampling is about 15 min. Even after exceeding the potlife of more than 100% these lacquer systems are still soluble. The reaction degree of the isocyanates is below 10% which is within the error range of the complete method. In order to test applicability of the method a sample is taken from the reaction batch before and after the processing. Each sample is derivatized directly. In the subsequent analysis the polyisocyanate concentrations in both samples should be identical within the tolerance range. In 1986 the analysis method was tested in practice in an extended investigation program [5].

10 References

[1] *Europäisches Komitee für Normung (CEN)* (1993) DIN EN 481 – Arbeitsplatzatmosphäre – Festlegung der Teilchengrößenverteilung zur Messung luftgetragener Partikel. Brüssel. Beuth Verlag, Berlin.
[2] *Tiesler A, Eben A* (1991) Hexamethylene diisocyanate (HDI), 2,4- and 2,6-toluylene diisocyanate (TDI; toluene-2,4-and 2,6-diisocyanate). In: Kettrup A (Ed) Analyses of hazardous substances in air, Vol. 1. VCH-Verlagsgesellschaft, Weinheim, 67–83.
[3] *Bayer AG* (1975) Anwendungstechnische Information, Order No KL 44270.
[4] *Deutsches Institut für Normung e.V. (DIN)* (1998) DIN EN ISO 11909, Bindemittel für Beschichtungsstoffe, allgemeine Prüfverfahren, Polyisocyanate. Beuth Verlag, Berlin.
[5] *Technischer Arbeitskreis der Fachabteilung Autoreparaturlacke im Verband der Lackindustrie e.V.* (1987) Konzentrationsmessungen in der Spritzkabinenluft – concentration measurements in spray booth atmospheres. Farbe Lack 11: 911–914.

Author: *M. Kuck*
Examiner: *J.U. Hahn*

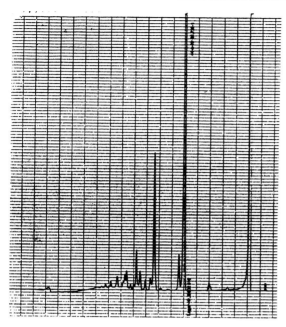

Fig. 1. Example of a chromatogram of polyisocyanate based on HDI.

Solvents
(continuously recording measurement of solvent vapours)

Method number   1

Application   Air analysis

Analytical principle Flame ionization detection

Completed in   June 1983

Summary

In the described method, organic gases and vapours are measured continuously by means of a flame ionization detector (FID). Unlike in gas chromatography, the substances are not separated in a column. The components contained in the sample air are recorded simultaneously but unspecifically and with different sensitivities.

By use of an in-built membrane pump, the sample air is drawn through a PTFE tube and a ceramic filter. The sample air flow is split so that a small flow is conducted to the hydrogen flame in the measuring cell. A DC voltage is applied between the flame tip burning and at the so called collector electrode. An ion current results as a measuring signal from the flame to the collector electrode if organic gases or vapours are present. It mainly depends on the number of carbon atoms in the molecule but the concentration is also important. The processed measuring signal is transmitted to a recorder. The recorded diagram shows the course of the concentration (concentration profile) in the sample air over the complete measuring period.

The indication of the measuring value is adjusted by use of a calibration gas containing a known concentration of propane in synthetic air.

For calibration defined concentrations of organic compounds (solvents) like gases or vapours are prepared in air and conducted into the apparatus. The relationship between the resulting measuring signal and the calibration gas concentration leads to the response factor. The response factor represents the ratio of the measuring signals which results at equal concentrations by volume of the analyte in the sample air and of propane in the calibration gas.

Precision:	Standard deviation (rel.)	$s = 0.5\%$
	Mean variation	$u = 1.0\%$
	referring to the final value of the measuring range	
Detection limit:	The detection limit depends on the hydrocarbon concentration in the ambient air (methane) or other organic compounds which is normally in the range of $1-2 \text{ mL/m}^3$	

Recommended flow rate in the sampling line:	> 5 L/min
Recommended flow velocity in the sampling line:	> 5 m/s
Recommended sampling time:	at least 1 h
Cross sensitivity:	Unspecific measuring method, all organic compounds are detected, however, with different sensitivity

Authors: *Th. zur Mühlen, L. Grupinski*
Examiner: *B. Striefler*

Solvents
(continuously recording measurement of solvent vapours)

Method number   1

Application   Air analysis

Analytical principle Flame ionization detection

Completed in   June 1983

Contents

1 General principles

In the described method, organic gases and vapours are measured continuously by means of a flame ionization detector (FID). Unlike in gas chromatography, the substances are not separated in a column. The components contained in the sample air are recorded simultaneously but unspecifically and with different sensitivities.

By use of an in-built membrane pump, the sample air is drawn through a PTFE tube and a ceramic filter. The sample air flow is split so that a small flow is conducted to the hydrogen flame in the measuring cell. A DC voltage is applied between the flame tip burning and at the so called collector electrode. An ion current results as a measuring signal from the flame to the collector electrode if organic gases or vapours are present. It mainly depends on the number of carbon atoms in the molecule but the concentration is also important. The processed measuring signal is transmitted to a recorder. The recorded diagram shows the course of the concentration (concentration profile) in the sample air over the complete measuring period.

The indication of the measuring value is adjusted by use of a calibration gas containing a known concentration of propane in synthetic air.

For calibration defined concentrations of organic compounds (solvents) like gases or vapours are prepared in air and conducted into the apparatus. The relationship between the resulting measuring signal and the calibration gas concentration leads to the response factor. The response factor represents the ratio of the measuring signals which results at equal concentrations by volume of the analyte in the sample air and of propane in the calibration gas.

2 Equipment, gases and chemicals

2.1 Equipment

Gas analyser with flame ionization detector, equipped with fittings for gases like hydrogen, synthetic air, zero gas, calibration gas and sample air, flow control and flowmeter for hydrogen and synthetic air, switching valve for zero gas, calibration gas or sample air, ceramic filter and membrane pump-both heatable, if necessary-split with back-pressure regulator and back-pressure gauge for the sample air, measuring cell (FID)-if necessary heatable-with ignition device, amplifier unit with different measuring ranges, measuring instrument and signal output (cf. Fig. 1)

Recorder, measuring range modulated with the signal output of the gas analyser, variable chart speed

PTFE tubing, internal diameter of about 4 mm, length up to 30 m

Gas collecting tubes with a lateral adapter and septum, volume of at least 1 L

10 μL Injection syringe for liquids

5 mL Gastight injection syringe

2.2 Gases

After-purified hydrogen, compressed gas cylinder
Synthetic air, compressed gas cylinder
Calibration gas: Propane in synthetic air (concentrations e. g. of about 500 mL/m^3 or
   about 10 mL/m^3 depending on the analytical problem), compressed gas cylinder

2.3 Chemicals

If the instrument has to be calibrated for individual solvents, the purity of the calibration substances has to be at least 99 %. Significant concentrations for example of stabilizers influence the accuracy of the determined response factors.

2.4 Additional equipment

The calibration (cf. Sect. 4.1) as well as the calculation of the mean value for the sampling can be carried out more easily by use of an electronic integrator.
In the calibration the response factors for the conversion of the calibration gas to the
used solvents can be calculated directly if a computing integrator for gas chromatography is available. For this purpose the apparatus should be equipped with the option of
an additional analysis run to serve as calibration and the option to set the sample
amount and default factors (cf. Sect. 4.2).
The synthetic air used as combustion air is suited to check the baseline before, during
and after the measurement. The synthetic air can be directly conducted into the zero
gas inlet of the instrument over a by-pass in the line of the combustion air and a pressure control connected behind.
Combustion air and zero gas in a sufficient purity can also be obtained from the outdoor air by catalytic after-burning. The so called zero gas generators with appropriate
equipment (catalyst, pump) are commercially available. They should always operate
with outdoor air. The use of the air at the workplace should strictly be avoided.
Tube fittings (e. g. Swagelok®) suitable for the sampling hose simplify the installation
of longer sampling lines.

3 Sampling and measurement

3.1 Setting of the instrument

The set up of the instrument and the installation of the sampling lines have to be chosen in a way that a hindrance or endangering of the employees at the workplace is ex-
cluded. Special care has to be taken of mobile parts of machines. In any case the set up
has to be out of areas of danger of explosion. Sampling lines up to a length of 30 m

can be installed without any problems. If necessary, the pumping capacity and the internal diameter of the sampling lines have to be enhanced in case of even longer sampling lines. For preheating the instrument is connected with the mains. The connections with the compressed gas cylinder for hydrogen, synthetic air, zero gas and calibration gas are installed and the operating pressures are adjusted at the pressure reducing valve according to the operating instructions. The switching valve is in the sampling position. After finishing the preheating and adjustment of the operating temperature the flame in the detector is ignited. The sampling line is connected with the instrument and the back pressure for the partial flow to the detector is controlled at the back-pressure gauge. If necessary, the back pressure is adjusted at the back-pressure regulator for example to the value as mentioned in the instruction manual. The chart speed of the paper is set at the recorder – a chart speed of 12 cm/h has proved suitable.

3.2 Checking of the instrument indication

Before the instrument indication is checked it has to be ensured that the compressed gas cylinders, especially those containing the calibration gas, are equilibrated at ambient temperature. The required measuring range is set and the zero gas (synthetic air) and the calibration gas are selected alternately using the same values feed of the back pressure. Possible deviations have to be adjusted by changing the pressure in the inlet pipe at the pressure reducing valve installed at the compressed gas cylinder. It is not permissible to change the setting at the back-pressure control.

By means of potentiometers the indications of the zero gas and the calibration gas are set to the required values. For measurements of solvent vapours or mixtures of solvent vapours at workplaces, an instrumental setting has proved efficient in which a full scale indication on the chart is corresponding with a propane concentration of 1000 mL/m^3. For the setting of the measuring value of the calibration gas the response factor has to be considered (cf. Sect. 5.1) if individual substances (e.g. trichloroethylene in the degreasing of metals) have to be determined.

3.3 Sampling

The reversing valve at the instrument is turned into the position sampling and the sample air is conducted over the sample line to the instrument. The measured values are recorded and integrated if possible. The instrument indication and the baseline are checked at regular intervals between the measurements and if necessary adjusted. It has to be checked also after finishing the measurement. Baseline deviations have to be considered in the calculation of the analytical results for each measuring period.

4 Calibration

For special applications stable calibration gases of known composition are commercially available in compressed gas cylinders (e. g. vinylchloride). Apart from this, the adjustment of the measuring value at the FID occurs by use of a calibration gas consisting of synthetic air under addition of a defined concentration of propane. In order to calibrate the instrument for a distinct organic compound (solvent) the ratio of the indication sensitivity between this compound and propane has to be determined. This proportional factor is defined as the response factor [1].

The calibration can be performed continuously or discontinuously. For the continuous mode of calibration gas mixing apparatus are suited. For this purpose a given constant concentration of solvent vapour in air is prepared in the laboratory and conducted into the gas analyser. The method of the discontinuously calibration is described as follows [2].

4.1 Calibration with the area evaluation method

The procedure to determine the response factors is specified in the following text. The operating parameters are set at the FID as described in Sect. 3.1. The chart speed at the recorder is at least 1 cm/s. A gas collecting tube is evacuated and filled with calibration gas (propane in synthetic air – cf. Sect. 2.2) taken from a compressed gas cylinder. The overpressure to the ambient air is balanced by a short opening of a cock. By use of a gastight syringe a defined volume ranging between 1 and 5 mL is taken through a septum from the gas collecting tube. It is injected into the apparatus and the resulting peak is recorded. This procedure is repeated several times. The number of possible injections without significant decrease of concentration in the gas collecting tube depends on the ratio between the dosage volume and the volume of the gas collecting tube. The injection into the injection port of the instrument has to be carried out in a rapid and regular way to avoid asymmetric peaks.

For the calibration of the FID different concentrations of vapour air mixtures of the selected compounds (solvents) are prepared in the gas collecting tube. These mixtures are injected into the gas analyser as mentioned before to record the resulting peaks. The concentrations should be in the same range as the calibration gas or lower. Calculated volumes of the liquids in the microliter range are injected through a septum into close gas collecting tubes containing air which is free from solvent vapours. At an ambient temperature of 20 °C the relationship between the dosed volume of the liquid and the volume concentration of a component C is as follows:

$$\sigma_C = \frac{V_C \cdot d_C}{M \cdot V_G} \cdot 24\,000 \ \frac{mL}{mole}$$

Legend:

σ_C Volume concentration of the component C in the gas collecting tube in mL/m^3
V_C Dosed volume of the liquid of the component C in μL

d_C Density of the liquid component C in g/mL
M Molecular weight of the component C in g/mole
V_G Volume of the gas collecting tube in L

The value of 24 000 mL/mole corresponds with the molecular volume of ideal gases at 20 °C (293 K) and 1013 hPa.

After the injection a residue-free vapourization and a homogenous mixing of the solvent with the air has to be awaited. In general this takes 20–30 minutes. In some cases it is recommended to fill some glass beads into the gas collecting tubes and to guarantee the homogeneous mixture by shaking.

The recorded peak areas of propane and the solvent are determined and each area is averaged. In case of symmetric peaks the area can be approximated by the product of the height and the width in the half of the peak height. The formula of Condal-Bosch [3] is applied in case of asymmetric peaks or peaks distorted by tailing. Then the average value of the peak width in 15 and 85% of the peak height is applied instead of the width at the half of the peak height (cf. Fig. 2).

The response factor is calculated as follows:

$$RF = \frac{\bar{F}_C \cdot \sigma_P}{\sigma_C \cdot \bar{F}_P}$$

Legend:

RF Response factor of the component C in contrast to the addition in the calibration gas (in this case propane)
σ_P Concentration by volume of propane in the calibration gas in mL/m^3
$\bar{F}_P$ Mean of the peak areas of the propane concentration σ_P in the calibration gas in cm^2 or arbitrary units
σ_C Given concentration of the component C in mL/m^3
$\bar{F}_C$ Mean of the peak areas of the concentration by volume σ_C of the component C in the same unit as $\bar{F}_P$

According to Sect. 5. 1 the calibration of a solvent vapour is obtained after adjusting the measuring value with propane by the division of the read value of propane through RF.

Remark: As the sensitivity of the indication is proportional to the number of the carbon atoms in the molecule (for aliphatic hydrocarbons) it is recommended to test the described method for calibration with pentane, hexane or heptane. Compared with propane the response factors have to be 1.67, 2.0 or 2.33.

4.2 Calibration by means of a computing integrator

The determination of the peak areas can be simplified by connecting a computing integrator with the FID. It is, however necessary to adapt the voltage output of the FID to the input of the integrator, for example by use of a change-over switch at the integrator. The adaption can also be carried out by use of a potential divider.

The response factor can be calculated directly if a computing integrator as it is commonly used in the gas chromatography is applied. The following procedure has proved to be suitable:

1. Before a series of calibration gas injections the integrator is started and it is stopped after the analysis run. The times from the beginning of the analysis run to the appearance of the peaks are printed as retention times with the corresponding values of the integrated areas.
2. The analysis is repeated as a calibration at the integrator according to the method of the external standard. For this purpose the calibration gas concentrations are feed for each retention time.
3. The obtained calibration factors are averaged, deleted in the integrator memory and the mean value is set as the default factor.
4. Before the series of injections of the solvent-vapour-air mixture the given concentration is set at the integrator as sample amount and the analysis run is started.
5. After the analysis run is finished the integrated value of each peak area is printed and the result corresponding with the retention time is calculated simultaneously by use of the default factor. If necessary the posistion of the decimal point has to be corrected. Response factors of about 0.3 (e. g. for methanol) to >3 (e. g. for benzene homologous compounds) can be expected.

5 Calculation of the analytical result

The measuring result obtained by means of the FID is a measuring diagram representing the course of the concentration of the solvent vapour during the measuring period. While at the measuring location only a pure solvent or preparation releasing vapours of a single compound (e. g. trichloroethylene) into the ambient air is applied, the diagram shows the concentration profile of the airborne vapour of this compound. The results represent sum parameters in case of mixtures of solvent vapours. They contain the individual components which deviate from their real part of the total concentration due to their different response factors.

5.1 Evaluation for individual components

Due to the adjustment of the measuring value at the FID during the measurement by use of a calibration gas and the response factor obtained by the calibration the concentration of the solvent vapour is calculated as follows:

$$\sigma_C = \frac{\Delta h_C \cdot \sigma_P}{\Delta h_P \cdot RF}$$

Legend:

σ_C Concentration by volume of the solvent C in mL/m^3
σ_P Concentration by volume of the addition (propane) in the calibration gas in mL/m^3

Δh_C Level of the measuring value of the component C in arbitrary units (e. g. scale units)

Δh_P Level of the measuring value of the addition (propane) in the calibration gas (in the same unit as for Δh_C)

RF Response factor of the solvent component referring to the addition (propane) in the calibration gas

There is another possibility to adjust the measuring value by use of the calibration gas. In this case a defined concentration of solvent vapour for example the maximum workplace concentration has to correspond with a given level of the measuring value on the recording chart.

$$\Delta h_P = \frac{\Delta h_{C,\text{Soll}} \cdot \sigma_P}{\sigma_C \cdot RF}$$

Legend:

Δh_P Level of the measuring value which has to be adjusted for the calibration gas (in arbitrary units, e. g. scale units)

$\Delta h_{C,\text{Soll}}$ Given level of the measuring value for the concentration by volume σ_C of the component C (in the same unit as indicated for Δh_P)

σ_P Concentration by volume of the calibration gas in mL/m^3

σ_C Given concentration by volume of the component C for the level of the measuring value $\Delta h_{C,\text{Soll}}$ in mL/m^3

RF   Response factor of the component C compared with the addition in the calibration gas

5.2 Evaluation of the measuring results in case of solvent vapour mixtures

In case of solvent vapour mixtures only an indication of the total concentration as equivalents of propane or as mg of hydrocarbons per m^3 is possible without further investigations.

More detailed indications are possible if the average composition of the solvent vapour is known from gas chromatographic analysis. For this purpose samples from the air measured by the analyser can be taken in gas collecting tubes by use of the bypass of the FID and investigated by means of gas chromatography. Even at strongly fluctuating total concentrations it can be shown that the ratio of the vapour concentrations of the individual components to each other is constant. Due to this constant composition a mean response factor of the actual mixture of the solvent vapours can be estimated from the response factors of the individual components. This leads to more definite explanation about the assignment between the measuring values and the total concentration of the solvent vapour (cf. Tab. 1).

Table 1. Calculation of an approximate average response factor ($\overline{RF}$) from the results of the gas chromatographic analysis of five air samples which taken by means of gas collecting tubes over a bypass at the FID.
(The solvent mixture consists of an alcohol-component F-acetic acid esters-D, E and G-as well as aliphatic and aromatic hydrocarbons A, B, C and H).

Air sample No	1		2		3		4		5	
Component	Response factor (RF)	Conc. x RF mL/m³		Conc. x RF mL/m³		Conc. x RF mL/m³		Conc. x RF mL/m³		Conc. x RF mL/m³
A	1.70	20	34	10	17	7	11.9	21	35.7	13	22.10
B	2.0	75	150	22	44	21	42	100	200	56	112
C	2.0	62	124	31	62	30	60	85	170	48	96
D	1.15	74	85.1	37	42.55	46	52.9	95	109.25	68	78.2
E	1.45	56	81.2	37	53.65	57	82.65	77	111.65	63	91.35
F	1.1	10	11	7	7.7	12	13.2	13	14.3	11	12.1
G	1.7	5	8.5	5	8.5	10	17	5	8.5	8	13.6
H	2.7	12	32.4	10	27	17	45.9	13	35.1	15	40.5
Sum		314	526.2	159	262.4	200	325.55	409	684.5	282	465.85
Quotient		1.68		1.65		1.63		1.67		1.65	
Mean ($\overline{RF}$)		1.66									

6 Reliability of the method

6.1 Instrumental parameters

The time until the operating temperature is reached and the measuring apparatus is stabilized (readiness for measurement) is about one hour. After this time every hour a baseline drift of less than 1% of the final value of the measuring range occurs. The 90-%-time is below one second. The dead time results from the length of the suction line and the flow rate.

6.2 Precision

The precision mainly depends on the exact adjustment and the stability of the gas flow rates (hydrogen, burning air, sample air). For the determination of the standard deviation calibration gas was charged into the provided inlet. The setting of the back pressure control was changed so that the normal setting and a quarter of this value was indicated at the back-pressure gauge. At each correct setting of the back pressure the measuring value was taken from the recording chart which was adjusted at 50% of the final value of the measuring range. A number of 20 measuring values obtained by this method lead to a relative standard deviation of 1% or a standard deviation of 0.5% of the final value of the measuring range.

6.3 Detection limit and sources of error

The detection limit is independent of the instrumental equipment but it depends on sources of error which occur from the background or the given content of hydrocarbons or other organic compounds in the ambient air.
The ubiquitous air concentration of methane is in the range of $1-2$ mL/m^3. It is influenced by the vegetation, the soil condition and other sources. The density of the traffic and industrial emissions are additional sources. This fact has to be considered in case of measurements in the lowest measuring range (about $0-10$ mL/m^3). To secure the measuring results a gas chromatographic investigation of the air samples is recommended.

7 Discussion

The described method has proved to be suitable for the continuous recording of pollutant profiles at workplaces. However, it has proved advantageous that the method is unspecific because of the manifold possibilities of applications, but in addition, it is necessary to perform gas chromatographic experiments to secure the results. If only one distinct compound is applied at workplaces the concentration can be read from the diagram. This is true for example for the degreasing and the purification of metals with trichloroethylene, tetrachloroethylene and 1,1,1-trichloroethane, in the processing of lacquers and other preparations which contain only one component of a solvent and – with limitations – in the processsing of polyester resins (interferences are possible due to cleaning works with acetone or dichloromethane). Even in those case a control by means of gas chromatography is absolutely necessary because unknown sources of error – leaking gas pipes and valves for natural gas or acetylene, gas operated radiation heaters, unobserved cleaning works – lead to wrong measuring results.
Quantitative measurements are strongly limited in the analysis of mixtures of solvent vapours. In any case the recording of the concentration profile always permits to detect the occurence and duration of enhanced concentrations and to make assignments to special working procedures. After the sampling (collecting phases, gas collecting tubes) the results of the measurements can often only be interpreted only by means of their concentration profiles. Under special conditions, estimations of concentrations in mixtures are possible by use of an average response factor (cf. Sect. 5.2).
The experiments to work out this method have been carried out by use of a flame ionization detector RS 5 from Ratfisch, München.

A computing integrator "Autolab System I" of Spectra Physics was used for the evaluation of the peak areas for the calibration and the calculation of the response factors.

A gas mixing apparatus of Telab-Labor und Technik, Gesellschaft für Meßtechnik in Duisburg was applied for the continuous preparation of the calibration gases of organic solvents. The calibration gases were obtainable from Linde AG, München and Messer Griesheim, Düsseldorf.

8 References

[1] *Dobson JG, Karas EL, Rooney TB* (1966) Anwendung von Flammen-Ionisations-Detektoren für die Analyse kontinuierlicher Prozesse. Arch. techn. Messen 361: 25–44.
[2] *Hinsch B, Rinne G, Reiser HJ* (1975) Experimenteller Vergleich zweier Meßverfahren zur Überwachung der Emission von Kohlenwasserstoffabgasen an Lackdraht-Einbrennöfen. Draht-Fachzeitschrift 26: 6.
[3] *Verein Deutscher Ingenieure (VDI)* (1995) VDI-Richtlinie 3481 Blatt 3, Messen gasförmiger Emissionen, Messen von flüchtigen organischen Verbindungen, insbesondere von Lösemitteln, mit dem Flammen-Ionisations-Detektor (FID). Beuth Verlag, Berlin.

Authors: *Th. zur Mühlen, L. Grupinski*
Examiner: *B. Striefler*

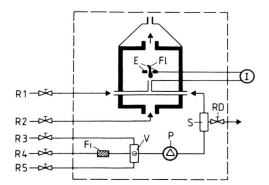

R1	= Hydrogen
R2	= Synthetic air
R3	= Calibration gas
R4	= Sample air
R5	= Zero gas
RD	= Back-pressure regulator
S	= Bypass
V	= Switching valve
P	= Pump
F_i	= Ceramic filter
E	= Electrode (collector electrode)
Fl	= Flame
I	= Ion current

Fig. 1. Gas analyser equipped with flame ionization detector.

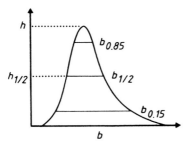

Fig. 2. Determination of the area of unsymmetric peaks according to the Condal-Bosch method [3]:

$$F = \frac{h \cdot (b_{0.15} + b_{0.85})}{2}$$

Styrene

Method number	1
Application	Air analysis
Analytical principle	Infrared spectrophotometry
Completed in	April 1992

Summary

Styrene absorbs infrared radiation at defined wavelengths. The concentration can directly be determined in the gas phase without sample preparation by measuring the extinctions of preselected absorption bands in a gas cuvette. For styrene it is suitable to measure the extinction at 11.1 μm ($\cong 900$ cm^{-1}).

The sample air is either directly drawn into the IR long-path gas cuvette (continuous measurement) or air samples are collected in gas collecting tubes or special gas collecting bags to be analysed later (discontinuous measurement).

For calibration defined concentrations of styrene are prepared in air atmosphere or in the gas cuvette, if possible. The extinctions are measured and plotted against the used concentrations to make the calibration curve.

The general experimental procedure is described in [1] "Infrared spectrophotometric determination of gases and vapours using long-path gas cuvettes".

Precision:	Standard deviation (rel.)	$s = 4\%$
	Mean variation	$u = 15\%$
	in a concentration range of 10–150 mL	
	styrene per m^3 air and $n = 3$ determinations	
Detection limit:	0.5 mL styrene per m^3 of air at an	
	optical path length of the gas cuvette of 21.75 m	
Sampling recommendation:	a) Continuous suction	
	b) 2 L Minimum volume in the gas collecting tube or gas collecting bag	

Styrene

$C_6H_5-CH=CH_2$

is a colourless, strongly light-refracting liquid with a characteristic odour. It is irritant to the eyes and to the mucuous membranes. The olfactory treshold limit is 0.20 mg/m^3 air. The currently valid MAK value (1998) is 20 mL/m^3 (86 mg/m^3) and styrene is listed in the peak limitation category II, 1 [2].

Styrene is used in numerous industrial fields. The working procedures are manifold and lead to different impacts in the work area. Styrene evaporates especially in the processing of unsaturated polyester resins (UP), e.g. in the manual lamination of mats (tank and boat manufacturing).

Authors: *L. Grupinski, W. Staub*
Examiners: *W. Wüstefeld, E. Bohne*

Styrene

Method number	1
Application	Air analysis
Analytical principle	Infrared spectrophotometry
Completed in	April 1992

Contents

1 General principles

Styrene absorbs infrared radiation at defined wavelengths. The concentration can directly be determined in the gas phase without sample preparation by measuring the extinctions of preselected absorption bands in a gas cuvette. For styrene it is suitable to measure the extinction at 11.1 μm ($\cong 900$ cm^{-1}).

The sample air is either directly drawn into the IR long-path gas cuvette (continuous measurement) or air samples are collected in gas collecting tubes or special gas collecting bags to be analysed later (discontinuous measurement).

For calibration defined concentrations of styrene are prepared in air atmosphere or in the gas cuvette, if possible. The extinctions are measured and plotted against the used concentrations to make the calibration curve.

The general experimental procedure is described in Reference [1] "Infrared Spectrophotometric Determination of Gases and Vapours using Long-path Gas cuvettes", see also [3].

2 Equipment and chemicals

2.1 Equipment

Infrared spectrophotometer, single or double-beam type, if possible equipped with a recording unit to scan the extinction at a constant wavelength

Long-path gas cuvette with variable path length and known volume

Vacuum pump (suction capacity about 1 m^3/h) to evacuate the gas collecting tubes and the gas cuvette, final pressure about 1 hPa

Sampling pump (flow rate of about 10 L/min) for sampling through pipes, for filling gas collecting tubes and gas collecting bags and for calibration.

Thermometer

Barometer

Gas collecting tubes for sampling, volume for example 2 L or 5 L, vacuum-tight and able to withstand implosions

Gas collecting bags for sampling, volume for example 10 L or 30 L (e.g. of Linde, München)

Gas collecting tube for calibration equipped with lateral adapter and septum, volume about 100 mL

Sampling tubing, e.g. of polytetrafluoroethylene (PTFE)

Tube fittings, e.g. of Viton, polypropylene, polyamide

Gastight syringes, e.g. 1, 5 and 10 µL volume

2.2 Chemicals

Styrene of known purity

Synthetic air

3 Sampling and sample transfer into the long-path gas cuvette

As described in Reference 1 the sampling can be carried out directly and continuously by use of a sampling pump or discontinuously by use of gas collecting tubes or gas collecting bags.
In addition to the transfer of the complete sample into the gas cuvette (continuous sampling or discontinuous sampling with gas collecting tubes and flushing with synthetic air) there is the possibility of a pressure equalization between the gas collecting tube and the evacuated gas cuvette (cf. [1]).

4 Analytical determination

The concentration of airborne styrene is determined at the wavelength 11.1 μm ($\cong 900$ cm^{-1}). The baseline method (cf. Fig. 3 in [1]) is applied for the optimum adjustment of the wavelength at the instrument. Especially in case of single-beam instruments an exact adjustment is of great importance. By use of a calibration gas (cf. Sect. 5 in [1]) the wavelength is adjusted to the absorption maximum of the selected band.
For the concentration of 10 mL styrene per m^3 air and more the optical path length in the cuvette is 21.75 m.

5 Calibration

The calibration is carried out according to Sect. 6 in [1].
For the calibration by direct injection a quantity of purified air is allowed to stream into the evacuated gas cuvette. Simultaneously a defined volume of styrene is injected. Calibration may also be carried out by injecting defined quantities of styrene into a gas collecting tube and by analyzing by means of infrared spectrophotometry.
The concentration of styrene in the gas cuvette is calculated according to the following equation:
Concentration in the gas cuvette (mL/m^3) =

$$\frac{\text{Injected volume of styrene } (\mu L) \cdot d\,(g/mL) \cdot V_m\,(L/mole)}{\text{Volume of the gas cuvette } (m^3) \cdot M\,(g/mole)}$$

Legend:

V_m 24.1 L (molecular volume at 20 °C and 1013 hPa)
M 104 g/mole (molecular weight of styrene)
d 0.9 g/mL (physical density)

The volumes as listed in Table 1 are injected into a gas cuvette of 5.62 L volume.

Table 1. Example for making a calibration curve.

Injected volume of styrene μL	Concentration in the gas cuvette mL/m^3
0.7	26
1.4	52
2.1	78
2.8	104

An injection of 1 μL styrene (liquid) into the gas cuvette of 5.62 L volume leads to a concentration in the cuvette of 37.1 mL/m^3 or 160.1 mg/m^3.
The calibration has to be checked before each measurement and depends on the apparatus used.
Additional calibration methods are described in Sect. 6 in [1].

6 Calculation of the analytical result

6.1 Continuous measurements

The concentration of styrene can directly be taken from the calibration curve.

6.2 Discontinuous measurements with gas collecting bags

If the volume of the gas collecting bag is greater than the volume of the gas cuvette the concentrations can directly be taken from the calibration curve. If not, this has to be considered in the calculation of the analytical result (cf. [1]).

6.3 Discontinuous measurements with gas collecting tubes

If the sample is introduced into the long-path gas cuvette by pressure equalization or if the volume of the gas collecting tube is smaller than the gas cuvette the styrene concentration in the gas cuvette is lower than the airborne styrene concentration at the measuring location. This has to be considered in the calculation of the analysis result as, mentioned in [1].

7 Reliability of the method

7.1 Precision

By use of a single beam instrument a relative standard deviation of $s = 4\%$ was determined from $n = 20$ individual measurements with concentrations up to 150 mL/m^3. This corresponds with a mean variation of $u = 15\%$.

7.2 Detection limit

The detection limit of styrene is 0.5 mL/m^3 air at an optical path length of the long-path gas cuvette of 21.75 m. However, it depends on the available instrument.

7.3 Sources of error

The solvents which are used for the cleaning of the machines and tools interfere with the determination of styrene at a wavelength of 11.1 μm ($\widehat{=} 900$ cm^{-1}).

8 Discussion

The method is reliably applied in the concentration range of 1–500 mL/m^3. The method is well suited to detect peak concentrations by means of the continuous measurement. A particular advantage of the method is the option to measure continuously as well as discontinuously. As an immediately indicating method it is suited for the testing of rehabilitation measures in work areas.

9 References

[1] *Grupinski L, zur Mühlen T* (1991) Infrared spectrophotometric determination of gases and vapours using long-path gas cuvettes. In: Kettrup A (Ed) Analyses of hazardous substances in air, Vol. 1. VCH-Verlagsgesellschaft, Weinheim, 25–39.
[2] *Deutsche Forschungsgemeinschaft (DFG)* (1998) List of MAK and BAT values 1998. Maximum concentrations and biological tolerance values at the workplace. Commission for the Investigation of Health Hazards of Chemical Compounds in the Work Area, Report No 34. WILEY-VCH Verlag GmbH, Weinheim.
[3] *Grupinski L* (1986) Messen von Gasen und Dämpfen mit IR-Spektrometrie und Langweg-Gasküvetten. Staub-Reinhalt Luft 46: 490–496.

Authors: *L. Grupinski, W. Staub*
Examiners: *W. Wüstefeld, E. Bohne*

Tetrahydrofuran

Method number   1

Application   Air analysis

Analytical principle   Gas chromatography

Completed in   April 1993

Summary

Measured air volumes from the breathing area are drawn through activated carbon tubes by means of a personal sampling pump. The adsorbed tetrahydrofuran (THF) is desorbed with 5 mL of dioxane. The gas chromatographic determination is performed by use of a flame ionization detector. The quantitative evaluation is carried out by a calibration curve. For this purpose the tetrahydrofuran concentrations of the calibration standards which are added with activated carbon are plotted against the peak areas calulated by an integrator.

Precision:	Standard deviation (rel.) $s = 2.4\%$
	Mean variation $u = 5.4\%$
	at an applied quantity of 12.54 mg
	and $n = 10$ samples
Detection limit:	1.27 mL/m^3 (equivalent with 3.75 mg/m^3)
	referring to a sample volume of 20 L
Recovery rate:	$\eta = 0.99$ (99%)
Recommended sampling time:	8 h
Recommended sample volume:	20 L

Tetrahydrofuran

Tetrahydrofuran (THF) is a colourless, inflammable liquid (molecular weight 72.11 g/mole, boiling point 66 °C) which is miscible with water, alcohol, ether and many common solvents in any ratio.

Tetrahydrofuran is commonly used in the lacquer and foil industry and it is also applied as a co-solvent for printing colours and adhesives. Due to its manifold reactions THF is also used for the manufacture of different reaction products.

The currently valid MAK value (1998) is 50 mL/m^3 or 150 mg/m^3 [1].

Authors: *E. Flammenkamp, A. Kettrup*
Examiners: *W. Krämer, W. Merz*

Tetrahydrofuran

Method number   1

Application   Air analysis

Analytical principle Gas chromatography

Completed in   April 1993

Contents

1 General principles

Measured air volumes from the breathing area are drawn through activated carbon tubes by means of a personal sampling pump. The adsorbed tetrahydrofuran (THF) is desorbed with 5 mL of dioxane. The gas chromatographic determination is performed by use of a flame ionization detector. The quantitative evaluation is carried out by a calibration curve. For this purpose the tetrahydrofuran concentrations of the calibration standards which are added with activated carbon are plotted against the peak areas calulated by an integrator.

2 Equipment, chemicals and calibration standards

2.1 Equipment

Gas chromatograph equipped with flame ionization detector
Recorder and integrator
Activated carbon adsorption tubes containing 600 mg activated carbon prepared from coconut shell, plastic caps for closing the tubes after the sampling, e. g. of MTC-GmbH, Müllheim
Personal air sampling pump, pumping capacity of about 2–5 L/h, e. g. of Compur, Model 4900 (equipped with a reading device)
Thermometer
Barometer
Sample vials equipped with PTFE coated septa and closure caps
Crimper
Glass cutter
20, 50, 100 and 200 mL Volumetric flasks
5, 20 and 25 mL Bulb pipette
2.5 mL Syringe for liquids
1 μL Syringe

2.2 Chemicals

Tetrahydrofuran, analytical grade, e. g. from Merck
Dioxane, analytical grade. e. g. from Merck

2.3 Calibration standards

A volume of 2.5 mL tetrahydrofuran is transferred into a 100 mL volumetric flask containing dioxane to prepare a stock solution. The flask is weighed and filled up to the mark with dioxane. By dilution with dioxane calibration standards are prepared from this stock solution containing about 1.4–55 mg tetrahydrofuran in 5 mL dioxane:

Table 1. Pipetting scheme for the preparation of calibration standards.

Volume of the stock solution mL	Final volume of the calibration standard mL	Concentration of the calibration standard mg/5 mL
25	50	55.5
25	100	27.8
15	100	16.7
5	50	11.1
5	100	5.6
2.5	100	2.8
2.5	200	1.4

3 Sample collection and preparation

The sample collection is carried out by use of a personal sampling pump and an adsorption tube filled with activated carbon over an 8-hour working day. The pump is connected with a reading unit and the number of the strokes is determined. The obtained sample volume is calulated as follows:

$$V = \frac{K \cdot N}{1000}$$

Legend:

V Volume in L
K Pump constant cm^3
N Number of the strokes

Temperature and ambient pressure are noted at the sampling location.
After the sampling the activated carbon tubes are immediately closed with plastic caps. They are opened for the sample preparation and the activated carbon is transferred each into vials which are then closed with septa. By use of a gastight syringe 5 mL of dixane are added through the septum. The samples are stored over night or shaked for 2 hours. In each series of analysis a new adsorption tube is prepared in the same way to check the blank value.

4 Operating conditions for gas chromatography

Column:	Material:	Fused silica
	Length:	25 m

	Stationary phase:	Ultra 2 (5% diphenylpolysiloxane	
		and 95% dimethylpolysiloxane)	
	Internal diameter	0.2 mm	
	Film thickness	0.33 m	
Detector:	Flame ionization detector		
Temperatures:	Column:	100 °C	
	Injector:	220 °C	
	Detector:	250 °C	
Carrier gas:	Helium:	Prepressure:	100 kPa
Detector gases:	Hydrogen:	Prepressure:	240 kPa
	Synthetic air:	Prepressure:	380 kPa
Injection volume:		1 µL	
	Split:	40 mL/min	
Analysis time:		5 min	

5 Analytical determination

The operating conditions are set as described in Sect. 4. Volumes of 1 µL each of the samples are injected at least twice. The evaluation is carried out by an integrator.

6 Calibration

5 mL each of the calibration standard as described in Sect. 2.3 are added with the activated carbon of an adsorption tube, stored over night and then analysed by gas chromatography. The measured peak areas are plotted against the used THF concentrations in 5 mL dioxane to make the calibration curve (cf. Fig. 1).

7 Calculation of the analytical result

By use of the obtained peak areas the corresponding weight of THF in mg is taken from the calibration curve. The corresponding concentration by weight (ρ) is calculated as follows:

$$\rho = \frac{X}{V \cdot \eta}$$

At 20 °C and 1013 hPa:

$$\rho_0 = \rho \, \frac{273 + t}{293} \cdot \frac{1013 \, \text{hPa}}{p}$$

The corresponding concentration by volume (σ) is (independent of pressure and temperature):

$$\sigma = \rho_0 \, \frac{24.1 \, \text{L} \cdot \text{mole}^{-1}}{72.11 \, \text{g} \cdot \text{mole}^{-1}} \cdot \frac{1013 \, \text{hPa}}{p} = \rho \cdot \frac{273 + t}{p} \cdot 1.155 \, \frac{\text{hPa} \cdot \text{mL}}{\text{mg}}$$

At $t = 20 \, °\text{C}$ and $p = 1013 \, \text{hPa}$:

$$\sigma = \rho \cdot 0.344 \, \frac{\text{mL}}{\text{mg}}$$

Legend:

X Concentration of THF by weight in mg per tube
V Sample volume in m^3
t Temperature of the ambient air in °C
p Pressure of the ambient air in hPa
ρ Airborne THF concentration by weight in mg/m^3 referring with t and p
ρ_0 Airborne THF concentration by weight in mg/m^3 at $t = 20 \, °\text{C}$ and $p = 1013 \, \text{hPa}$
σ Airborne THF concentration in mL/m^3
η Recovery rate

8 Reliability of the method

8.1 Precision

The precision of the method including the sample preparation was determined by use of a dynamic calibration gas. At a flow rate of 22.7 mL/min each activated carbon tube was loaded with calibration gas for six minutes. The concentration was 92 µg/mL THF in synthetic air. The precision was determined from ten sampling procedures. The relative standard deviation was $s = 2.4\%$ and the mean variation was $u = 5.4\%$.

8.2 Recovery rate

The same experimental arrangement was used to check the deposition efficiency of THF on the activated carbon. An average amount of 99% of the applied THF (cf. Table 2) was recovered at sampling times between 3 and 30 min. On tubes connected behind the test tubes THF could not be detected by a calibration gas (80 % relative air humidity) up to a used quantity of 60 mg.

Table 2. Determination of the recovery rate.

mg THF applied	mg THF recovered	Recovery rate %
6.3	6.1	97.4
12.5	12.3	97.8
25.1	26.6	106.1
37.6	38.5	102.2
62.7	63.3	101.1

8.3 Detection limit

Under the operating conditions described above and a sample volume of 20 L the detection limit is 1.27 mL/m^3 (3.75 mg/m^3).

8.4 Specificity

Under the operating conditions described above the following substances do not interfere with the method: Ethanol, diethyl ether, acetone, dichloromethane, carbon disulfide, cyclohexane and benzylalcohol.
For special separation tasks the column temperature has to be changed or a column of a different polarity has to be used.

9 Discussion

The described method permits an exact determination of airborne THF concentrations in the range of a tenth up to a multifold of the currently valid MAK value of 150 mg/m^3.
Loaded tubes (closed with caps) can be stored up to three weeks at room temperature without loss of substance.
The recovery rate and the precision were also determined by loading the tubes with a calibration gas (corresponding with 1000 ppm) for 30 minutes. The precision was only a bit lower than the values described here.

Apparatus: Gas chromatograph F 22 (Perkin-Elmer) equipped with flame ionization detector

10 References

[1] *Deutsche Forschungsgemeinschaft (DFG)* (1998) List of MAK and BAT values 1998. Maximum concentrations and biological tolerance values at the workplace. Commission for the Investigation of Health Hazards of Chemical Compounds in the Work Area, Report No 34. WILEY-VCH Verlag GmbH, Weinheim.

Authors: *E. Flammenkamp, A. Kettrup*
Examiner: *W. Krämer, W. Merz*

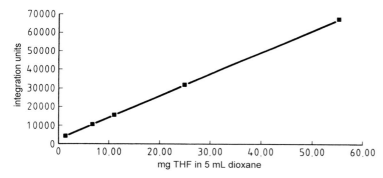

Fig. 1. Example of a calibration curve for the gas chromatographic determination of airborne tetrahydrofuran.

Tetrahydrothiophene (Thiolane)

Method number   1

Application   Air analysis

Analytical principle Gas chromatography

Completed in   June 1982

Summary

The airborne vapours of tetrahydrothiophene are absorbed in *n*-heptane and determined by a gas chromatograph equipped with a flame photometric detector. The quantitative evaluation is carried out by a calibration curve in which the tetrahydrothiophene concentrations of the used calibration standards in a double logarithmic measure are plotted against the peak heights.

Precision:	Standard deviation (rel.) $s = 3.3–3.5\%$
	Mean variation $u = 6.9–7.3\%$
	in a concentration range of $6.1–15.7$ µg/mL
	tetrahydrothiophene in *n*-heptane and $n = 20$ gas
	chromatographic determinations
Detection limit:	0.27 mL/m^3 tetrahydrothiophene in air (equivalent
	with 1.0 mg/m^3) referring to a sample volume of 10 L
Recovery rate:	$\eta = 0.98$ (98%)
Recommended sampling time:	20 min
Recommended sample volume:	10 L

Tetrahydrothiophene (Thiolane)

$$H_2C\!-\!CH_2$$
$$H_2C \quad CH_2$$
$$\diagdown\,S\,\diagup$$

Tetrahydrothiophene (Thiolane) is a colourless liquid (molecular weight 88 g/mole, boiling point 119 °C at 1013 hPa, vapour pressure 19.3 hPa at 20 °C, density 0.999 g/mL) with an intensive smell and a strong irritant effect to the mucuous membranes. It is used as a natural gas odorant.

Author: *A. Eben*
Examiner: *K. Wrabetz*

Tetrahydrothiophene (Thiolane)

Method number	1
Application	Air analysis
Analytical principle	Gas chromatography
Completed in	June 1982

Contents

1 General principles

The airborne vapours of tetrahydrothiophene are absorbed in n-heptane and determined by a gas chromatograph equipped with a flame photometric detector. The quantitative evaluation is carried out by a calibration curve in which the tetrahydrothiophene concentrations of the used calibration standards in a double logarithmic measure are plotted against the peak heights.

2 Equipment, chemicals and solutions

2.1 Equipment (cf. Fig. 1)

Gas chromatograph equipped with a flame photometric detector (FPD), a sulfur speci-
fic filter (393 nm) and a 1 mV recorder
Gas wash bottles (50 mL volume) equipped with frits G_1 (W_1, W_2)
Thermos bottles (B_1, B_2)
Pump, pumping capacity of at least 1 L/min (P)
Throttle valve (V_1)
Calibrated gasmeter suitable for volume measurements at a flow rate of 1 L/min (G)
Thermometer (T_g, T_a)
Barometer (B_a)
10, 20, 50, 100 mL Volumetric flasks
5 mL Measuring pipettes

2.2 Chemicals

Tetrahydrothiophene (thiolane) of known purity degree
n-Heptane, (Uvasol® grade)

2.3 Calibration standards

A quantity of 36.25 mg tetrahydrothiophene for example is weighed into a 100 mL volumetric flask containing 10 mL heptane (cf. Fig. 2). The flask is filled up to the mark
with n-heptane. 2.5 mL of this solution are transferred into a 50 mL volumetric flask
and diluted with n-heptane to the mark. From this stock solution (containing 18.12 µg/
mL for example) calibration standards are prepared by dilution with heptane as listed
below:

Table 1. Pipetting scheme for the preparation of calibration standards.

Volume of the stock solution mL	Final volume of the calibration standard mL	Concentration of the calibration standard µg/mL
5.0	10.0	9.06
5.0	20.0	4.53
2.5	20.0	2.27

3 Sample collection and preparation

Two serially connected gas wash bottles each containing 6 mL *n*-heptane are cooled with ice (in a thermos bottle) and fitted with the sampling device (cf. Fig. 1). A flow rate of 0.5 L/min is adjusted by use of the throttle valve. The applied sample volume (V_Z), the temperature of the gasmeter (t_g) and the ambient temperature (t_a) are noted as well as the ambient pressure.

After finishing the sample collection the content of both wash bottles W_1 and W_2 is transferred each into a 10 mL volumetric flask. The bottles are rinsed with *n*-heptane and diluted to the mark. The concentration is determined by gas chromatography.

4 Operating conditions for gas chromatography

Column:	Material:	Steel
	Length:	1.8 m
	Internal diameter:	2.2 mm (1/8 inch tube)
Stationary phase:	5% Silicone oil DC 550 on Chromosorb G	
	(AW-DMCS), 80–100 mesh	
Detector:	Flame photometric detector equipped with sulfur specific filter	
	393 nm	
Temperatures:	Column:	140 °C isothermal
	Injector block:	150 °C
	Detector:	200 °C
Carrier gas:	Helium	(10 mL/min)
Burning gases:	Hydrogen	(50 mL/min)
	Air	(40 mL/min)
	Oxygen	(10 mL/min)
Injection volume:	1.0 μL	

5 Analytical determination

The operating conditions are adjusted at the gas chromatograph. Volumes of 1 μL are taken from the wash bottles W_1 and W_2 and several times injected into the injector block. To carry out the quantitative determination volumes of each 1 μL from at least three calibration standards of different tetrahydrothiophene concentrations are injected.

6 Calibration

Using the measured peak heights of the calibration standards a calibration curve is drawn in which the peak heights are plotted against the used concentrations ($\mu g/mL$) in a double-logarithmic measure. An example of a calibration curve is shown in Fig. 2. Chomatograms of different tetrahydrothiophene concentrations are shown in Fig. 3.

7 Calculation of the analytical result

The peak height of the analyte is determined and the corresponding concentration is read off from the calibration curve. This value is multiplied with the volume of the sample solution (mL) to obtain the weight of tetrahydrothiophene (μg) in the sample solution. The sum of the weights from both wash bottles represents the complete weight X (μg) of tetrahydrothiophene.

The corresponding concentration ρ (mg tetrathydrothiophene per m^3 ambient air) is calculated according to the following equation:

$$\rho = \frac{X}{V_Z \cdot \eta} \cdot \frac{273 + t_g}{273 + t_a}$$

At 20 °C and 1013 hPa:

$$\rho_0 = \rho \, \frac{(273 + t_a)}{293} \cdot \frac{1013 \, hPa}{p_a}$$

The corresponding concentration by volume σ – independent of pressure and temperature is:

$$\sigma = \rho_0 \, \frac{24.1 \, L \cdot mole^{-1}}{88.0 \, g \cdot mole^{-1}} = \rho \cdot \frac{273 + t_a}{p_a} \cdot \frac{1013 \, hPa}{293} \cdot \frac{24.1 \, L \cdot mole^{-1}}{88.0 \, g \cdot mole^{-1}}$$

$$\sigma = \rho \cdot \frac{273 + t_a}{p_a} \, 0.947 \, \frac{hPa \cdot mL}{mg}$$

At $t_a = 20\,°C$ and $p_a = 1013$ hPa:

$$\sigma = \rho \cdot 0.274 \, \frac{mL}{mg}$$

Legend:

X	Sum of the tetrathiophene weights from W_1 and W_2 in μg
ρ	Concentration by weight in mg/m^3 at t_a and p_a (see below)
ρ_0	Concentration by weight in mg/m^3 at 20 °C and 1013 hPa

σ Concentration by volume in mL/m^3
V_Z Sample volume in L
t_g Temperature in the gasmeter in °C
t_a Temperature of the ambient air in °C
p_a Pressure of the ambient air in hPa
η Recovery rate

8 Reliability of the method

8.1 Precision

Each of 20 single determinations of tetrahydrothiphene solutions in *n*-heptane containing average concentrations of 6.1 and 15.7 µg/mL yielded standard deviations (rel.) of
$s = 3.3$ and 3.5%. The mean variation u was 6.9 and 7.3%.

8.2 Recovery rate

The completeness of the absorption of tetrahydrothiophene vapours in *n*-heptane could
be tested by the experimental arrangement as shown in Fig. 4. A defined mass of the analyte was dissolved in 0.1–0.5 mL acetone, pipetted into the wash bottle D and warmed in
a water bath to 25–40 °C. The vapours were drawn through two serial gas wash bottles
W_1 and W_2 (equipped with frits) applying a flow rate of about 0.5 L/min of dry air. An
amount of 95–100% of the used tetrahydrothiophene was recovered in the first wash bottle (cf. Tab. 2). In cases tetrahydrothiophene could be detected in the second wash bottle
the concentration was below 0.1% and therefore could be neglected.

Table 2. Recovery rate of tetrahydrothiophene.

Applied weight of tetrahydrothiophene	Recovered	
mg	mg	%
187.6	178	94.9
173.4*	168*	96.9*
125.0	125	100.0
130.0*	130*	100.0*
83.4	80	95.9
41.7	41.0	98.3
43.3*	43.3*	100.0*

* Results of the examiner

$$\bar{M} = \frac{M\,(\%)}{n} = 98.0\%$$

8.3 Detection limit

Under the operating conditions as mentioned the detection limit of tetrahydrothiophene is 0.27 mL/m^3 (corresponding with 1.0 mg/m^3) at a sample volume of 10 L air.

8.4 Sources of error

Under the operating conditions as mentioned the following compounds in concentrations up to 1 mg/mL do not interfere with the method: Acetone, benzene, trichloromethane, diethyl ether, dichloromethane, N,N-dimethylformamide, 1,4-dioxane, ethyl acetate, 2-methoxyethanol, methanol, n-hexane, isopropanol, methylisobutylketone, n-pentane, toluene, xylene, tetrachloromethane, 1,2-dichloroethane, tetrahydrofuran.

9 Discussion

The described method permits a rapid and exact determination of airborne tetrahydrothiophene.

A gas chromatograph 5730 A equipped with an FPD, a sulfur specific Filter 393 nm and a 1 mV recorder of Hewlett Packard was used to work out this method.

Applying the operating conditions as mentioned the retention time of tetrahydrothiophene was 40 s. In order to confirm the method the results were tested by use of a gas chromatograph 3700 equipped with an FPD of Varian. Under the same operating conditions the retention time was 1.3 min at a column temperature of 110 °C. In this checking the peak areas were plotted against the concentrations of the calibration standards. It could be shown that the relationship was not linear in all cases. This fact is possibly depending on the used apparatus.

Author: *A. Eben*
Examiner: *K. Wrabetz*

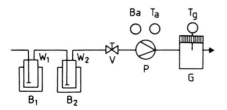

W$_1$, W$_2$	Gas wash bottles equipped with frits
B$_1$, B$_2$	Thermos bottles filled with ice water
V	Throttle valve
P	Pump
G	Gasmeter
T$_g$	Thermometer (sampling air)
T$_a$	Thermometer (ambient air)
Ba	Barometer

Fig. 1. Sampling apparatus.

peak height
mm

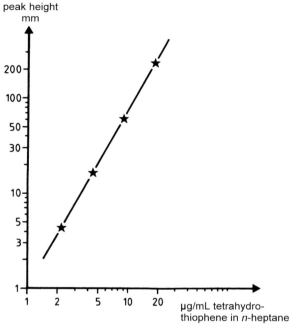

200 —

100 —

50 —

30 —

10 —

5 —

3 —

1 —

1   2   5   10   20   µg/mL tetrahydro-
                            thiophene in *n*-heptane

Fig. 2. Example of a calibration curve for gas chromatographic determination of airborne tetra-hydrothiophene concentrations.

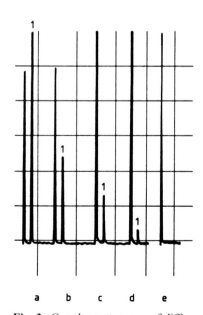

a = 18.12 µg/mL *n*-heptane (atten. 8)
b = 9.06 µg/mL *n*-heptane (atten. 8)
c = 4.53 µg/mL *n*-heptane (atten. 4)
d = 2.27 µg/mL *n*-heptane (atten. 4)
e = *n*-Heptane (Atten. 4)
1 = Tetrahydrothiophene peak

Retention time of tetrahydrothiophene: 40 s

a   b   c   d   e

Fig. 3. Gas chromatograms of different concentrations of tetrahydrothiophene.

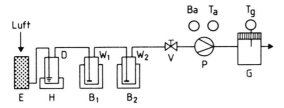

Fig. 4. Apparatus to test the recovery rate.

E	Drying tower (CaCl$_2$)
D	50 mL Wash bottle (containing tetrahydrothiophene)
H	Water bath
W$_1$, W$_2$	Gas wash bottles equipped with frits
B$_1$, B$_2$	Thermos bottles filled with ice water
V	Throttle valve
P	Pump
G	Gasmeter
T$_a$, T$_g$	Thermometer
Ba	Barometer

1,1,1-Trichloroethane

Method number   1

Application   Air analysis

Analytical principle Gas chromatography

Completed in   April 1993

Summary

Using a sampling pump measured air volumes are drawn through activated carbon in adsorption tubes. The adsorbed 1,1,1-trichloroethane is desorbed with carbon disulfide and then analysed in a gas chromatograph equipped with a flame ionization detector. For the quantitative evaluation a calibration curve is used in which the 1,1,1,-trichloroethane concentrations of the calibration standards are plotted against the peak areas obtained with an integrator.

Precision: Standard deviation (rel.) $s = 1.9\%$ and 3.6%
                    Mean variation $u = 4.9\%$ and 9.3%
                    for $n = 6$ determinations and concentrations of
                    4100 mg/m^3 and 100 mg/m^3
Quantification limit: 0.25 mg/m^3 (for a sample volume of 40 L)
Recovery rate: $\eta = 0.98$ (98%)
Recommended sampling time: 2 h
Recommended sample volume: 40 L

1,1,1-Trichloroethane

CH_3-CCl_3

1,1,1-Trichloroethane is a colourless liquid with a pleasant ether-like odour (molecular weight 133.42 g/mole, melting point $-32\,°C$, density 1.349 g/mL); it is insoluble in water but soluble in common organic solvents. 1,1,1-Trichloroethane is non-flammable. The currently valid MAK value (1998) is 200 mL/m^3 (1100 mg/m^3) [1]. Higher vapour concentrations cause mucosal irritation, headaches and balance disorders. Because 1,1,1-trichloroethane defats the skin, skin contact can result in rashes. Hepatic and renal damage is possible.

1,1,1-Trichloroethane is mainly used as a solvent in dry cleaning and for cold and hot cleaning and degreasing of metal workpieces.

Authors: *N. Lichtenstein, M. Bernhards*
Examiners: *T. zur Mühlen, R. Töpfer*

1,1,1-Trichloroethane

Method number	1
Application	Air analysis
Analytical principle	Gas chromatography
Completed in	April 1993

Contents

1 General principles

Measured air volumes are drawn through activated carbon in adsorption tubes with a sampling pump. The adsorbed 1,1,1-trichloroethane is desorbed with carbon disulfide and then analysed in a gas chromatograph equipped with a flame ionization detector. For the quantitative evaluation a calibration curve is used in which the 1,1,1,-trichloroethane concentrations of the calibration standards are plotted against the peak areas obtained with an integrator.

2 Equipment, chemicals and solutions

2.1 Equipment

Gas chromatograph equipped with flame ionization detector (FID)
Integrator
Activated carbon adsorption tubes (for example type B from Dräger)
Pump, flow rate 20–25 L/h (for example type Alpha 1 from DuPont)
Gasmeter
10, 25, 100 and 500 µL Injection syringes for liquids
Glass vials with screw caps and washers (15 mL)
5 mL Pipettes
5 mL Volumetric flasks

2.2 Chemicals

Carbon disulfide, analytical grade
1,1,1-Trichloroethane, purity >97%

2.3 Calibration standards

To prepare the stock solution about 400 mg 1,1,1-trichloroethane is weighed exactly and transferred into a 5 mL volumetric flask. The flask is filled up to the mark with carbon disulfide. Then from this stock solution the volumes listed in the following table are pipetted into 5 mL volumetric flasks and diluted up to the marks with carbon disulfide.

Table 1. Pipetting scheme for the preparation of calibration standards.

Volume of the stock solution µL	Concentration of the calibration standard mg/5 mL	Corresponding airborne concentration for a sample volume of 40 L air mg/m^3
20	1.6	40
200	16	400
400	32	800
800	64	1600
1200	96	2400
3600	288	7200

3 Sample collection and preparation

Using a flow stabilized personal sampling pump or a pump controlled by a gasmeter, the sample of air is drawn through activated carbon in an adsorption tube at a flow rate of 20 L/h (for a maximum sampling time of 2 h). For a sampling time of 8 hours the flow rate has to be reduced to 4 L/h. If necessary, two activated carbon adsorption tubes have to be connected in series (relative air humidity >80%; higher temperatures).

The loaded tubes are closed with plastic caps, labelled and the sampling data are recorded. If possible, the tubes should be stored in a cool place. The loaded collecting phases can be stored for 2 weeks without losses. For desorption of the 1,1,1-trichloroethane, the activated carbon is transferred into a 15 mL glass vial and 5 mL carbon disulfide is added. After 2 hours the desorption is complete.

4 Operating conditions for gas chromatography

Column:	50 m OV 1 (internal diameter 0.32mm, film thickness 30 µm)	
Split:	1:80	
Detector:	Flame ionization detector	
Temperatures:	Oven:	50 °C isothermal, 8 min
		50 °C → 80 °C (5 °C/min)
		80 °C isothermal, 5 min
		80 °C → 180 °C (5 °C/min)
	Injector block: 200 °C	
	Detector: 230 °C	
Carrier gas:	Helium (60 kPa)	
Detector gas:	Hydrogen: 34 mL/min	
	Synthetic air: 520 mL/min	
	(Parameters for the instrument used: "Sichromat", Siemens)	
Injection volume:	1 µL	

The above temperature program is appropriate for samples containing additional solvent components. Otherwise it should be adapted to the actual application.

5 Analytical determination

The gas chromatographic operating conditions are set as described in Sect. 4. Volumes of 1 µL of each of the solutions to be analysed are injected. The evaluation is carried out with an integrator.

6 Calibration

Volumes of 1 µL (cf. Sect. 2.3) are taken from each of the calibration standards and analysed by gas chromatography (cf. Fig. 1). The calibration function is determined from the peak areas (integrator).
As the calibration curve is linear in the concentration range of 5–1000 mg/m^3 (for 40 L air) a single-point calibration may also be performed but then the calibration concentration should be in the same concentration range as the samples.

7 Calculation of the analytical result

The 1,1,1-trichloroethane concentrations (mg/5 mL CS$_2$) corresponding to the peak areas (peak heights) obtained are read from the calibration curve. Given the same elution solution volumes the value obtained corresponds with the weight of 1,1,1-trichloroethane in the sample in mg (X).
The corresponding concentration ρ (mg 1,1,1-trichloroethane in per m^3 of air) is calculated according to the following equation:

$$\rho = \frac{X}{V_Z \cdot \eta} \cdot \frac{273 + t_g}{273 + t_a}$$

And for 20 °C and 1013 hPa:

$$\rho_0 = \rho \, \frac{273 + t_a}{293} \cdot \frac{1013 \, \text{hPa}}{p_a}$$

The corresponding concentration in ml/m^3 σ – independent of pressure and temperature – is given by:

$$\sigma = \rho_0 \, \frac{24.1 \, \text{L/mole}}{133.4 \, \text{g/mole}} = \rho \cdot \frac{273 + t_a}{p_a} \cdot \frac{1013 \, \text{hPa}}{293} \cdot \frac{24.1 \, \text{L/mole}}{133.4 \, \text{g/mole}}$$

$$\sigma = \rho \cdot \frac{273 + t_a}{p_a} \cdot 0.625 \, \frac{\text{hPa} \cdot \text{mL}}{\text{mg}}$$

And for $t_a = 20\,°C$ and $p_a = 1013$ hPa:

$$\sigma = \rho \cdot 0.181 \, \frac{\text{mL}}{\text{mg}}$$

Legend:

X Weight of 1,1,1-trichloroethane in the extraction solution in mg
V_Z Measured volume of the air sample in m^3

η Recovery rate
t_g Temperature in the gasmeter in °C
t_a Temperature of the ambient air in °C
p_a Atmospheric pressure in hPa
ρ Airborne 1,1,1-trichloroethane concentration in mg/m^3
          for t_a and p_a as described above
ρ_0 Airborne 1,1,1-trichloroethane concentration in mg/m^3
          at 20 °C and 1013 hPa
σ Airborne 1,1,1-trichloroethane concentration in mL/m^3

8 Reliability of the method

8.1 Precision

1,1,1-Trichloroethane concentrations of 100 and 4100 mg/m^3 in air were generated in a dynamic calibration gas apparatus to determine the precision of the method. Six samples at each concentration were taken, prepared and analysed as described. The resulting standard deviations were 3.6% (100 mg/m^3) and 1.9% (4100 mg/m^3) with mean variations of 9.3% and 4.9%.

8.2 Recovery rate

In the concentration range examined, the recovery rate is 98% ($\eta = 0.98$). It is independent of the concentration.

8.3 Quantification limit

Under the conditions described above the quantification limit is 1.9 µg 1,1,1-trichloroethane per mL solution. This corresponds to a concentration of 0.25 mg/m^3 for a sampled air volume of 40 L.

8.4 Specificity

Compounds with the same retention time as 1,1,1-trichloroethane may interfere with this determination. To exclude interferences, a second column of different polarity can be used.

9 Discussion

With this method, both personal and workplace-typical analytical data can be obtained. The sampling device is small, mobile and free of liquids.
The method described permits accurate determination of airborne 1,1,1-trichloroethane concentrations in the range of 5 mg/m^3 up to twice the currently valid MAK value. The recovery rate in the tested concentration range is constant.
If possible, the loaded adsorption tubes should be kept cool. They may be stored for two weeks without losses. Two tubes have to be connected in series if the relative humidity is higher than 80%.

10 References

[1] *Deutsche Forschungsgemeinschaft (DFG)* (1998) List of MAK and BAT values 1998. Maximum concentrations and biological tolerance values at the workplace. Commission for the Investigation of Health Hazards of Chemical Compounds in the Work Area, Report No 34. WILEY-VCH Verlag GmbH, Weinheim.

Authors: *N. Lichtenstein, M. Bernhards*
Examiners: *T. zur Mühlen, R. Töpfer*

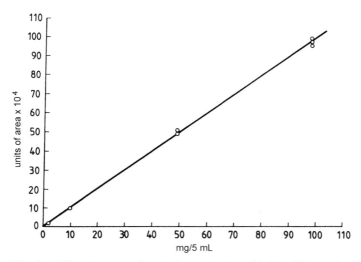

Fig. 1. Calibration curve for the determination of 1,1,1-trichloroethane.

Urea Herbicides (Metoxuron, Monuron, Monolinuron, Metobromuron, Diuron, Isoproturon, Linuron, Neburon)

Method number   1

Application   Air analysis

Analytical principle High performance liquid chromatography

Completed in   June 1992

Summary

The method described here is simple and rapid and permits the simultaneous determination of a series of urea herbicides. For this purpose measured air volumes are drawn through silica gel adsorption tubes. The adsorbed urea herbicides are eluted with methanol and determined in a high performance liquid chromatograph equipped with a UV detector. Calibration standards are used for the quantitative evaluation.

Precision:	Standard deviation (rel.) $s = 0.1$ and 3.5%
	Mean variation $u = 0.3$ and 9%
	at concentrations of 2.0 and 50.0 µg urea herbicides per m^3 air and for $n = 10$ determinations
Quantification limit:	0.05 mg urea herbicides per m^3 air for a sample volume of 10 L air and a final volume of 1.0 mL
Recovery rate:	$\eta = 1.0$ (100%) for a 50 µg sample of urea herbicides
	$\eta = 0.94-1.0$ (94–100%) for a 2 µg sample of urea herbicides
Recommended sampling time:	30 min–6 h
Recommended sample volume:	1–10 L
	For quantitative sample collection of particle-bound herbicides the flow rate has to be kept constant at 1.25 m/s $\pm$ 10%.

Metoxuron (3-(3-chloro-4-methoxy-phenyl)-1,1-dimethylurea

$$CH_3O-\underset{Cl}{\overset{}{\bigcirc}}-NH-CO-N\overset{CH_3}{\underset{CH_3}{\diagdown}}$$

Metoxuron is a colourless, crystalline substance (molecular weight 228.68 g/mole, melting point 126–127 °C). It is a selective herbicide for pre-emergence and post emergence treatment.

Monuron (3-(4-chlorophenyl)-1,1-dimethylurea)

$$Cl-\bigcirc-NH-\underset{\underset{O}{\|}}{C}-N\overset{CH_3}{\underset{CH_3}{\diagdown}}$$

Monuron is a colourless, crystalline substance (molecular weight 198.66 g/mole, melting point 174–175 °C). It is a root herbicide used as a pre-emergence as well as a general herbicide (photosynthesis inhibitor).

Monolinuron (3-(4-chlorophenyl)-1-methoxy-1-methylurea)

$$Cl-\bigcirc-NH-\underset{\underset{O}{\|}}{C}-N\overset{OCH_3}{\underset{CH_3}{\diagdown}}$$

Monolinuron is a colourless, crystalline substance (molecular weight 214.65 g/mole, melting point 79–80 °C). It is a selective pre-emergence herbicide and also a photosynthesis inhibitor.

Metobromuron (3-(4-bromophenyl)-1-methoxy-1-methylurea)

$$Br-\bigcirc-NH-\underset{\underset{O}{\|}}{C}-N\overset{OCH_3}{\underset{CH_3}{\diagdown}}$$

Metobromuron is a colourless, crystalline substance (molecular weight 259.11 g/mole, melting point 95–96 °C). It serves as a selective pre-emergence herbicide.

Diuron (3-(3,4-dichlorophenyl)-1,1-dimethylurea)

Diuron is a colourless, crystalline substance (molecular weight 233.10 g/mole, melting point 158–159 °C). It is a pre-emergence root herbicide.

Isoproturon (3-(4-isopropylphenyl)-1,1-dimethylurea)

Isoproturon is a colourless, crystalline substance (molecular weight 206.29 g/mole, melting point 152–153 °C). It is a selective pre-emergence and post-emergence herbicide.

Linuron (3-(3,4-dichlorophenyl)-1-methoxy-1-methylurea)

Linuron is a colourless, crystalline substance (molecular weight 249.10 g/mole, melting point 93–94 °C). It is used as a selective herbicide.

Neburon (3-(3,4-dichlorophenyl)-1-methyl-1-n-butylurea)

Neburon is a colourless, crystalline substance (molecular weight 275.18 g/mole, melting point 101.5–103 °C). It is used as a selective root herbicide.

Author: *A. Eben*
Examiner: *A. Kettrup*

Urea Herbicides (Metoxuron, Monuron, Monolinuron, Metobromuron, Diuron, Isoproturon, Linuron, Neburon)

Method number   1

Application   Air analysis

Analytical principle   High performance liquid chromatography

Completed in   June 1992

Contents

1 General principles

With sampling pump, measured air volumes are drawn through silica gel adsorption tubes. The adsorbed urea herbicides can be eluted with methanol. They are determined by high performance liquid chromatography. Calibration standards are used for the quantitative evaluation.

2 Equipment, chemicals and solutions

2.1 Equipment

Sampling pump for personal sampling (e. g. from Auer, Type C 210)
Universal battery-charging set (e. g. from Auer)
Stand for the adsorption tubes (e. g. from Auer)
Silica gel adsorption tubes (Dräger, type G)
High performance liquid chromatograph equipped with UV detector (247 nm)
Column: length 20 cm, internal diameter 2.1 mm, stationary phase: ODS Hypersil, 5 μm
Rotary evaporator
Thermometer
Barometer
100 mL Round bottom flasks
1, 5, 25 and 50 mL Volumetric flasks
Funnel (5.5 cm diameter)
1 mL Bulb pipettes

2.2 Chemicals

Metoxuron, 99% purity
Monuron, 99% purity
Monolinuron, 99% purity
Metobromuron, 99% purity
Diuron, 99% purity
Isoproturon, 99% purity
Linuron, 99% purity
Neburon, 99% purity
(These standard substances are available, e. g. from Riedel de Haen)
Methanol for chromatography or residue analysis
Orthophosphoric acid, 85%, analytical grade
Triethylamine

2.3 Solutions

Buffer solution for HPLC:
8.0 mL of triethylamine and 4.0 mL of orthophosphoric acid are pipetted into a 1 L volumetric flask and diluted up to the mark with deionized water.

2.4 Calibration standards

Stock solution:
A stock solution containing a mixture of the urea herbicides to be determined is prepared. 10 mg of each individual compound is weighed exactly in a weighing bottle, transferred into a 50 mL volumetric flask and diluted up to the mark with methanol. This solution contains 0.2 mg urea herbicide per mL. By dilution with methanol the calibration standards are prepared from this stock solution (cf. Tab. 1).

Table 1. Pipetting scheme for the preparation of calibration standards.

Volume of the stock solution mL	Final volume of the calibration standards mL	Concentration of the individual herbicides in the solution µg/mL
2.0	20.0	20.0
1.0	20.0	10.0
0.5	20.0	5.0
0.1	10.0	2.0
0.1	20.0	1.0
0.05	20.0	0.5

3 Sample collection and preparation

Using a pump for personal sampling, measured air volumes (1–5 L air/30 min or 10 L air/6 hours) are drawn through silica gel adsorption tubes. If the flow rate is also maintained at 1.25 m/s ±10% (inspiration rate) particulate herbicides may also be determined quantitatively.
After sampling the tube outlet is connected to a funnel and the silica gel is eluted with about 50 mL of methanol. The eluate is then concentrated at 30 °C in a rotary evaporator to a volume of about 0.5 mL. Depending on the expected concentration, it is transferred into a 1 mL volumetric flask and diluted up to the mark with methanol. If necessary, the solution is diluted again (e. g. 1:5). In each analysis series a fresh adsorption tube is processed in the same way to determine the blank value. Adsorption tubes which cannot be processed immediately after sampling must be closed with plastic caps.

4 Operating conditions for high performance liquid chromatography

Instrumental conditions:
Column   Material: Steel
                          Length: 20 cm
                          Internal diameter: 2.1 mm
Stationary phase: ODS Hypersil
Detector: UV detector wave length: 247 nm
Column temperature: Room temperature
Mobile phase: Eluent A: Buffer solution (cf. Sect. 2.3)
                          Eluent B: Methanol
Gradient program: Start: 65% A and 35% B
                                              changing to 75% B within 30 min
Flow rate: 0.35 mL/min
Injection volume: 5 μL
Retention times: Metoxuron: 6.0 min
                          Monuron: 7.6 min
                          Monolinuron: 10.6 min
                          Metobromuron: 12.0 min
                          Diuron: 13.6 min
                          Isoproturon: 13.9 min
                          Linuron: 16.7 min
                          Neburon: 24.1 min

5 Analytical determination

The HPLC apparatus is set up as described in Sect. 4. Volumes of 5 μL each of the solutions to be analysed were injected three times sequentially. For calibration, calibration
standards containing the urea herbicides in concentrations like those in the samples are
injected repeatedly between the samples to be analysed (cf. Sect. 2.4).

6 Calibration

To check the linearity of the analysis function, the calibration standards (cf. Sect. 2.4)
are analysed by means of HPLC (cf. Fig. 3). The measured peak areas were plotted
against the concentrations of the solutions injected. The calibration curves were linear
in the concentration range of 0.5–20.0 μg/mL methanol (cf. Fig. 1 and 2).
The evaluation of the analytical data was carried out using calibration standards.

7 Calculation of the analytical result

The peak areas obtained for the calibration standard and the analysed solution are calculated by the integrator. The weight of the urea herbicide (X in μg) in the air sample is calculated according to the following equation:

$$X = \frac{A \cdot C}{B} \cdot F$$

The concentration ρ (mg urea herbicide/m^3 of air) is given by:

$$\rho = \frac{X}{V_Z \cdot \eta} \cdot \frac{273 + t_g}{273 + t_a}$$

At 20 °C and 1013 hPa:

$$\rho_0 = \rho \, \frac{273 + t_a}{293} \cdot \frac{1013 \, \text{hPa}}{p_a}$$

Legend:

A Peak area obtained for the analysis solution
B Peak area obtained for the calibration standards
C Concentration of the calibration standards in μg/mL
F Dilution factor
V_Z Volume of air drawn through the adsorption tube in L
X Weight of the urea herbicide in μg in the air sample
η Recovery rate
t_g Temperature in the gasmeter in °C
t_a Temperature at the sampling location
p_a Atmospheric pressure at the sampling location in hPa
ρ Airborne urea herbicide concentration in mg/m^3 at temperatures t_a and p_a
ρ_0 Airborne urea herbicide concentration in mg/m^3 at 20 °C and 1013 hPa

8 Reliability of the method

8.1 Stability and shelf life

The behaviour of the active agents on the surface of the adsorbent was investigated by loading silica gel adsorption tubes with a mixture of the urea herbicides to be determined (50 μg of each of the individual components/100 μL methanol) and connecting the loaded tube to a sampling pump. 10 L of air were drawn through the tubes at a flow rate of 0.167 L/min. After sampling the tubes were closed with plastic caps and stored at ambient temperature. The elution of the tubes and the analytical determination of the methanolic solutions were carried out immediately, after 20 hours and after three days. A decrease in concentration was never observed.

8.2 Precision

The precision was determined from the recovery rate and applies for the whole method. For applied substance weights of 2.0 and 50.0 µg the values of the standard deviations (rel.) s and those of the mean variations u obtained in 5 tests are shown in Table 2.

Table 2. Standard deviation (rel.) s and mean variation u.

Urea herbicide	2.0 µg		50 µg	
	Standard deviation (rel.) s %	Mean variation u %	Standard deviation (rel.) s %	Mean variation u %
Metoxuron	3.5	9.0	1.6	4.1
Monuron	1.4	3.6	0.2	0.5
Monolinuron	1.4	3.6	0.5	1.3
Metobromuron	1.4	3.6	0.3	0.8
Diuron	1.7	4.4	0.1	0.3
Isoproturon	1.5	3.9	0.1	0.3
Linuron	2.1	5.4	0.3	0.8
Neburon	1.7	4.4	0.3	0.8

8.3 Recovery rate

To determine the recovery, methanolic solutions of the urea herbicides were prepared. The concentration of the individual components was 20 µg/mL (solution 1) and 500 µg/mL (solution 2). Each of 5 adsorption tubes (silica gel type G) were loaded with 100 µL of solution 1 (2.0 µg absolute weight) and another 5 with 100 µL of solution 2 (50 µg absolute weight). The tubes were connected to a sampling pump. At a flow rate of 0.167 L/min 10 L of air was drawn through the tubes. After sampling, the silica gel tubes containing the higher concentration of the herbicides were eluted with 50 mL of methanol. The eluate was concentrated to 5.0 mL. For the tubes containing the lower herbicide concentration the eluate was 50 mL and the final volume 1.0 mL (after concentration in a rotary evaporator).

The results are summarized in Tables 3 and 4. The recovery rates were in the range of 0.94–1.0 (94 and 100%).

In an additional experiment in which the two silica gel layers from the tube were eluted separately the herbicides were found only in the first layer.

Table 3. Recovery of the urea herbicides from adsorbent loaded with a herbicide mixture (50.0 µg absolute for each component).

Urea herbicide		Observed concentration (Prepared concentration: 50 µg absolute)					Average recovery rate
Metoxuron	µg	50.56	51.56	49.35	50.61	50.82	
	%	101.1	103.1	98.7	101.2	101.6	1.01
Monuron	µg	50.64	50.56	50.41	50.61	50.51	
	%	101.3	101.1	100.8	101.2	101.0	1.01
Monolinuron	µg	50.24	49.86	49.66	50.11	50.18	
	%	100.5	99.7	99.3	100.2	100.4	1.00
Metobromuron	µg	50.39	50.19	50.09	50.42	50.32	
	%	100.8	100.4	100.2	100.8	100.6	1.01
Diuron	µg	50.34	50.24	50.32	50.42	50.29	
	%	100.7	100.5	100.6	100.8	100.6	1.01
Isoproturon	µg	50.37	50.32	50.28	50.45	50.29	
	%	100.7	100.6	100.6	100.9	100.6	1.01
Linuron	µg	50.48	50.31	50.26	50.63	50.47	
	%	101.0	100.6	100.5	101.3	100.9	1.01
Neburon	µg	50.39	50.31	50.48	50.57	50.62	
	%	100.8	100.6	101.0	101.1	101.2	1.01

Table 4. Recovery of the urea herbicides from adsorbent loaded with a herbicide mixture (2.0 µg absolute for each component).

Urea herbicide		Observed concentration (Prepared concentration: 2.0 µg absolute)					Average recovery rate
Metoxuron	µg	1.78	1.89	1.85	1.96	1.89	
	%	89.0	94.5	92.5	98.0	94.5	0.94
Monuron	µg	2.02	1.99	1.97	2.02	1.96	
	%	101.0	99.5	98.5	101.0	98.0	0.97
Monolinuron	µg	1.96	1.99	1.95	1.92	1.93	
	%	98.9	99.5	97.5	96.0	96.5	0.98
Metobromuron	µg	2.01	2.03	1.99	1.98	1.96	
	%	100.5	101.5	99.5	99.0	98.0	1.00
Diuron	µg	2.01	2.01	1.99	2.00	1.93	
	%	100.5	100.5	99.5	99.9	96.5	0.99
Isoproturon	µg	2.01	2.02	1.99	1.98	1.94	
	%	100.5	101.0	99.5	99.0	97.0	0.99
Linuron	µg	2.01	2.01	1.98	1.94	1.92	
	%	100.5	100.5	99.0	97.0	96.0	0.99
Neburon	µg	2.06	2.03	2.03	2.08	1.99	
	%	103.0	102.5	101.5	104.0	99.5	1.02

8.4 Specificity

Also investigated were some substances which occur together with the urea herbicides in commercial products and therefore can be sprayed together with the herbicides. The given substances were added to the mixture of urea herbicides as well as to the individual components (concentration: 10 µg/mL).

Interference with the determination of the urea herbicides was not found with the following substances:

Amitrole (3-amino-1,2,4-triazole)
Anilazine (2,4-dichloro-6-(2-chloroanilino)-1,3,5-triazine)
Bromacil (5-bromo-3-*sec*-butyl-6-methyluracil)
Dalapon (2,2-dichloropropionic acid)
Methabenzthiazuron (1,3-dimethyl-3-(2-benzthiazolyl)-urea)
Paraquat (1,1'-dimethyl-4,4'-bipyridine dichloride)
Simazine (2-chloro-4,6-bis-ethylamino-*s*-triazine)

At higher concentrations atrazine ((2-chloro-4-ethylamino-6-isopropylamino)-1,3,5-triazine) can interfere with the determination of diuron. Accurate evaluation of the peak areas is possible up to concentrations of 10 µg/mL methanol.

8.5 Quantification limit

0.05 mg urea herbicide per m^3 air can be detected in a sample volume of 10 L of air with an elution volume of 50 mL and a final volume of 1.0 mL of solution.

9 Discussion

8 herbicides can be determined simultaneously and quantitatively by means of the rapid and simple method described here. Because urea herbicides may also occur bound to particles, for the sampling procedure it makes sense to maintain a suction velocity of 1.25 m/s $\pm$ 10 % (inspiration rate). For the adsorption from air, not only silica gel type G but also silica gel type B can be used. Desorption tests have shown that elution with methanol dropwise through the tubes gives better results than the elution of the silica gel in a beaker. No loss of herbicides was observed during concentration of the eluate.

Apparatus: High performance liquid chromatograph HP 1090 equipped with work station, UV detector (DAD) and an autosampler from Hewlett-Packard.

Author: *A. Eben*
Examiner: *A. Kettrup*

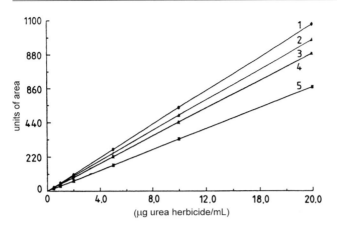

Fig. 1. Calibration curves for metoxuron (1), monuron (2), monolinuron (3), metobromuron (4), linuron (5). (For (3) and (4) the data were identical.)

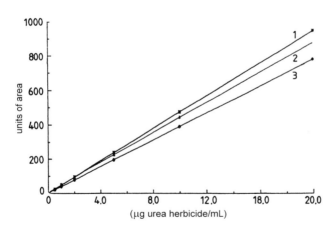

Fig. 2. Calibration curves for diuron (1), isoproturon (2) and neburon (3).

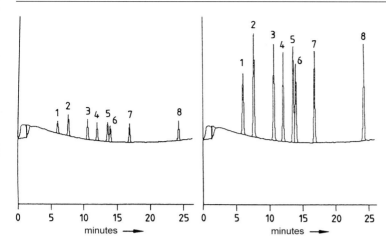

Fig. 3. Liquid chromatograms of the urea herbicides.

a. Metoxuron   1,0 µg/mL   b. Metoxuron   5 µg/mL
   Monuron   1.0 µg/mL   Monuron   5 µg/mL
   Monolinuron   1.0 µg/mL   Monolinuron   5 µg/mL
   Metobromuron   1.0 µg/mL   Metobromuron   5 µg/mL
   Diuron   1.0 µg/mL   Diuron   5 µg/mL
   Isoproturon   1.0 µg/mL   Isoproturon   5 µg/mL
   Linuron   1.0 µg/mL   Linuron   5 µg/mL
   Neburon   1.0 µg/mL   Neburon   5 µg/mL

Members and Guests of the Working Subgroup

Analyses of Hazardous Substances in Air of the Commission of the Deutsche Forschungsgemeinschaft for the Investigation of Health Hazards of Chemical Compounds in the Work Area

Leader:

Prof. Dr. A. Kettrup
GSF-Forschungszentrum Neuherberg
Institut für Ökologische Chemie
Ingolstädter Landstraße 1
85764 Oberschleißheim

Members:

Prof. Dr. J. Angerer
Institut für Arbeits-, Sozial- und Umweltmedizin
der Universität Erlangen-Nürnberg
Schillerstraße 25/29
91054 Erlangen

Dr. M. Blaszkewicz
Institut für Arbeitsphysiologie
ZE Analytische Chemie
Ardeystraße 67
44139 Dortmund

Dr. R. Heinrich-Ramm
Zentralinstitut für Arbeitsmedizin
Adolph-Schönfelder-Straße 5
22083 Hamburg

Priv.-Doz. Dr. Dr. U. Knecht
Institut und Poliklinik für Arbeits- und Sozialmedizin
der Universität Gießen
Aulweg 129/III
35392 Gießen

Prof. Dr. W. Riepe
Universität Salzburg
Institut für Chemie und Biochemie
Hellbrunnerstraße 34
A-5020 Salzburg

Guests:

Dr. M. Ball
ERGO-Forschungsgesellschaft mbH
Geierstraße 1
22305 Hamburg

Dr. D. Breuer
Berufsgenossenschaftliches Institut
für Arbeitssicherheit – BIA
Alte Heerstraße 111
53754 Sankt Augustin

Dr. H. G. Gielen
Landesamt für Umweltschutz und
Gewerbeaufsicht Rheinland-Pfalz
Rheinallee 97–101
55118 Mainz

Dr. U. Giese
Deutsches Institut für Kautschuktechnologie
Eupenerstraße 33
30519 Hannover

Dr. K. Goßler
Landesgewerbeanstalt Bayern
Analytikzentrum
Tillystraße 2
90431 Nürnberg

Dr. Habarta
Bayerisches Landesamt für Arbeitsmedizin
und Sicherheitstechnik
Pfarrstraße 3
80538 München

Dr. J. U. Hahn
Berufsgenossenschaftliches Institut
für Arbeitssicherheit – BIA
Alte Heerstraße 111
53754 Sankt Augustin

Dr. R. Hebisch
Bundesanstalt für Arbeitsschutz und Arbeitsmedizin
Friedrich-Henkel-Weg 1–25
44149 Dortmund

Dipl.-Ing. M. Hennig
Berufsgenossenschaftliches Institut
für Arbeitssicherheit – BIA
Alte Heerstraße 111
53754 Sankt Augustin

Dr. W. Kleiböhmer
Institut für Chemo- und Biosensorik e.V.
Mendelstraße 7
48149 Münster

Dr. W. Krämer
BASF AG
Labor für Umweltanalytik
DUU/OU-Z 570
67056 Ludwigshafen

Dr. M. Kuck
BAYER AG
ZF-DAL Geb. 013
51368 Leverkusen

Dr. N. Lichtenstein
Berufsgenossenschaftliches Institut
für Arbeitssicherheit – BIA
Alte Heerstraße 111
53754 Sankt Augustin

Dr. C. P. Maschmeier
Landesamt für Arbeitsschutz des Landes Sachsen-Anhalt
Brauereistraße 21 a
06846 Dessau

Dipl.-Ing. K. H. Pannwitz
Drägerwerk AG
Moislinger Allee 53/55
23542 Lübeck

Dipl.-Ing. K. Riegner
BAYER AG
PF-Zentrum Monheim, Geb. 6660
51368 Leverkusen

Dr. U. Risse
Klinik und Poliklinik
für Dermatologie/Allergologie
am Biederstein
Biedersteiner Straße 29
80802 München

Dipl.-Chem. U. Schröter
Amt für Arbeitsschutz
Marckmannstraße 129b
20539 Hamburg

Dr. I. Stanetzek
Hessisches Landesamt für Umwelt
Zentralstelle für Arbeitsschutz
Ludwig-Mond-Straße 33
34121 Kassel

Dipl.-Ing. M. Tschickardt
Landesamt für Umweltschutz und
Gewerbeaufsicht Rheinland-Pfalz
Rheinallee 97–101
55118 Mainz

Dipl.-Ing. K. Zuber
Wehrwissenschaftliches Institut für Werk-,
Explosiv- und Betriebsstoffe (WIWEB)
Landshuterstraße 70
85435 Erding

Secretariat:

Dr. J. Gündel
Institut und Poliklinik für Arbeits-, Sozial- und Umweltmedizin
der Universität Erlangen-Nürnberg
Schillerstraße 25/29
91054 Erlangen

Secretariat of the commission:

Dr. P. Marth
GSF-Forschungszentrum Neuherberg
Institut für Ökologische Chemie
Ingolstädter Landstraße 1
85764 Oberschleißheim

Dr. C. Pohlenz-Michel
GSF-Forschungszentrum Neuherberg
Institut für Toxikologie
Ingolstädter Landstraße 1
85764 Oberschleißheim

Contents of Volumes 1–3

DATE DUE

DEMCO 38-297